CO-OPERATIVE CREDIT FOR AGRICULTURAL SECTOR

CO-OPERATIVE CREDIT FOR AGRICULTURAL SECTOR

Dr. N. RADHAKRISHNAN, M.A., M.Phil., Ph.D.,
Lecturer in Economics
Muthurangam Government Arts College
Vellore - 632 002, Tamil Nadu

SERIALS PUBLICATIONS
NEW DELHI

First Published - 2006

ISBN: 81-8387-032-5

Published by

SERIALS PUBLICATIONS
4830/24, Prahlad Street, Ansari Road
Darya Ganj, New Delhi-110002 (India)
Phone : 23245225
Fax : 91-11-23272135
E-mail: serials@satyam.net.in

CONTENTS

PREFACE

Agriculture is the oldest occupation in the world and it continues to be the largest even today. Agriculture assumes greater importance in developing country like India for its varied contributions to the National Economy. When one looks at Indian Economy whether from the point of accelerating the process of income generation, widening the scope of employments, solving the problem of balance of payment situation, one has to turn to agriculture and its development.

Sustained growth of agriculture is a pre-requisite for the overall economic development of our country, which depends upon, among other things, mainly availability of adequate credit facility. The provision of timely and adequate credit facilities is a necessary imperative. The tempo of agricultural development could be accelerated only through the provision of adequate credit to the farmers at the right time, at the right place and through the right institution.

A farmer's income fluctuates as agriculture is a 'gamble in monsoon' and is liable for drought, floods, excessive rainfall, pests, cattle disease, fluctuating prices and yields. The farmer's problems is one of seasonal production cycle. While his needs are non-seasonal. Agriculture has two uncertainty, viz. production uncertainties and price uncertainty, where as industrialists face only price uncertainty.

The present system of agricultural credit in India is grouped under two headings, viz., the institutional and non-institutional. The non-institutional factors consists mainly of the professional and agricultural money-lenders, landlords, Traders, Commission agents, Relatives and Friends. The institutional factors comprises of Government, Co-operatives, Commercial banks and Regional Rural Banks. In spite of the significant expansion of the institutional factors, non-institutional agencies continue to play a dominant role in rural credit.

"Credit supports the farmers as the hangman's rope supports the hanged." Agricultural credit is a problem, when it cannot be obtained and it is also a problem when it can be obtained. In India it is a two-fold problem of inadequacy and unsuitability that is being continually faced by the farmers. The farmers are further handicapped from getting cheap credit because they are mostly individualistic and suspicious of combining with each other. This has retarded the availability of cheap finance.

The Co-operative credit system suffers from certain serious weaknesses at present. The continuous existence of weak central co-operative Banks has also been retarding the progress of co-operative credit movement in the country. The weak central co-operative banks will not be able to provide the required refinance facilities to the PACB. Another major problem of Co-operative financing in India has been that of mounting over-dues. The higher over-dues has affected the overall financial health of the rural credit structure. In the above both causes central co-operative Banks are ineligible for availing of refinance facility from National Bank for Agricultural and Rural Development (NABARD).

The gap between the total members and the actual loan users are widening, the over-dues are mounting and many societies are put in dormitory.

It is significant to note that agricultural credit provided by the co-operatives is much more significant in quantum than that provided by the other institutional agencies. The co-operative system is the most suitable institutional agency for this purpose in the conditions of the Indian villages. This book assures greater significance as it is said that the agriculture at present is in deep crisis and the agrarian population is in acute debt crisis.

I have evaluated the existing rural credit system in the light of this concept and have suggested suitable measures for removal of its deficiencies. This study is based on primary data through interview and also various secondary sources of information. This book is contains eight chapters.

This work has grown out of my doctoral dissertation submitted to the University of Madras in December 2002, under the supervision of Dr. P. Govindarajan, Professor and Head of the Department of Economics (Retd.), Presidency college, Chennai-5. I am heavily indebted to him for not only initiating me into research but also took pains in taking care and constantly encouraged me to complete my doctoral research successfully and in time.

My Friend and colleagues at the college and outside who helped me directly or indirectly in the preparation of this Book. I am indeed, grateful to all of them.

I must thank S.K. Jha Managing Director of Serials Publications, New Delhi who took keen interest in publication of this book.

Last but not the least, I would thank my parents, wife and daughters for their support and constant source of inspiration to complete this work.

(N. RADHAKRISHNAN)

LIST OF TABLES

LIST OF MAPS

LIST OF ABBREVIATIONS

Agri. Labour	- Agriculture Labour
ARC	- Agricultural Refinance Corporation
CBs	- Commercial Banks
CRAFICARD	- Credit for Agriculture and Rural Development
DPAP	- Drought prone area programme
Govt.	- Government
IRDP	- Integrated Rural development Programme
KPACB	- Kalathure Primary Agricultural Co-operative Bank
LAL	- Landless Agricultural Labour
LDB	- Land Development Bank
LF	- Large famers
LIC	- Life Insurance Corporation of India
LT	- Long-term
MAF	- Mutual Assistance Fund
MF	- Marginal farmers
MT	- Medium term
MTC	- Medium term conversion
NABARD	- The National Bank for Agricultural and Rural Development Bank
NPAs	- Non-Perfoming Assets
OD	- Overdue
PACBs	- Primary Agricultural Co-operative Banks
PMRY	- Prime Minister's Rozgar Yojana
RBI	- Reserve Bank of India
RCCIs	- Rural Credit Co-operative Institutions
RD A/C	- Recurring Deposits Account

RFI	-	Rural Financial Institutions
RRBs	-	Regional Rural Banks
SB A/C	-	Savings Account
SCBs	-	State Co-operative Banks
SF	-	Small farmers
SFDA	-	Small farmers Development Agency
SHGs	-	Self-Help Groups
SSI	-	Small-Scale Industries
ST	-	Short-term
TIIC	-	Tamil Nadu Industrial Investment Corporation
TNSC Bank	-	Tamil Nadu State Co-operative Bank
TPACB	-	Thamaraipakkam Primary Agricultural Co-operative Bank
VDCCB	-	Vellore District Central Co-operative Bank

Chapter - 1

INTRODUCTION

Agriculture is the oldest occupation in the world and it continues to be the largest even today. The majority of the population of the developing world, probably nearly two-thirds of the total, are dependent upon it for a living[1]. Agriculture occupies a key position in Indian Economy. Sustained growth of agriculture is a pre-condition for the overall economic development of our country, which depends upon, among other things, availability of adequate credit facility[2]. Agriculture, the primary occupation of man kind, fulfils the basic requirement of food since the emergence of human race[3].

Agriculture assumes greater importance in developing country like India for its varied contributions to the National Economy. Even today, it contributes 30 percent of the total National Income and 67 percent of the total population depends on this sector for their livelihood[4]. When one looks at Indian economy whether from the point of view of accelerating the process of income generation, or widening the scope of employment or solving the problem of balance of payments situation, one has to turn to agriculture and its development[5].

The recent breakthrough in agriculture production has, no doubt, made it possible to achieve spectacular increase in productivity. However, this breakthrough has also pushed up the credit needs of the farm sector. In order to sustain and intensify the use of new inputs in agriculture, provision of timely and adequate credit facilities is a necessary imperative. The tempo of agricultural development could be accelerated only through the provision of adequate credit to the farmers at the right time, at the right place and through the right institution.

Making credit available to farmers in adequate amounts and at reasonable costs, however, does not imply, that agricultural investments should be especially favoured by capital gifts (or) heavily subsidised loans. This can result in a general misallocation of resources both within agriculture and also between the agricultural and non-agricultural sectors of economy. But it does imply that institutional arrangements should as far as possible overcome the characteristic obstacles which agriculture presents to the provision of credit at low costs. In India, it has been found that private agencies like moneylenders cannot be fitted into any scheme of rural credit meant to serve the needs of rural development. Hence, efforts are being made to provide institutional agencies for rural credit[6].

Credit is a vital factor in farm production in Indian agriculture where the prospect for plough back is weak as compared to trade and industry. The vast number of small and marginal farmers constitute the bulk of agricultural population at the subsistence level. They are virtually without surplus (or) savings left for future investment and production purposes. After harvest they possess very little surplus for further consumption after repayment of the credit. A farmer's income fluctuates as agriculture is a 'gamble in monsoon' and is liable for drought, floods, excessive rainfall, pests, cattle disease, fluctuating prices and yields. The farmer's problem is one of seasonal production cycle, while his needs are non-seasonal[7]. Agriculture has two uncertainty viz: production uncertainties and price uncertainty, whereas industrialists face only price uncertainty. Farmers seek credit not only for production purposes but also to finance basic consumption needs and to repay debts[8].

The agricultural community consists of large, medium, semi-medium, small, marginal farmers and agricultural labourers. Among the agriculturists small and marginal farmers form a major segment. Small farmers operate at a subsistence level and are always in the grip of a vicious circle. The problem of adequate and cheap credit is one of the perennial problems of Indian agriculture[9]. Adequate and timely credit is a crucial constituent of any programme of agricultural development. Provision of good credit facility to the farmer at the right time, in the right place and in the right proportion is a requisite for the transformation of traditional agriculture.

Farm credit may be classified on the basis of period, purpose, security-given, creditor-wise and borrower-wise. Based on the length of time it may also be classified as (a) Short-term, (b) Medium-term and (c) Long-term credits.

The provision of adequate funds on reasonable terms is an important constituent of any strategy of rural development. The exorbitant high rate of interest charged by the moneylenders, the oligopolistic nature of their business as well as other malpractices associated with them create insurmountable difficulties to the farmers. The problem of finding an alternative to moneylenders has been under consideration for quite some time and the Co-operative Credit Societies Act was passed as early as 1904. Even after this the entire rural sector depended mostly on the non-institutional agencies like moneylenders, commission agents and traders for their requirements. The co-operatives were no doubt in the field, but the extent of financial support extended by them has been limited in relation to the needs[10].

Agriculture cannot develop unless credit facilities are available to the peasantry in areas which are not viable. The supply of credit to such areas will be highly effective in rehabilitating the agriculturists. The need for such credit has become all the more important in the context of new strategies of farming and the introduction of high yielding varieties. No doubt there are various agencies to finance the farmers, but in India the role of co-operatives cannot be underestimated. For progressive agriculture adequate agricultural credit facility is most essential[11].

The All-India Rural Credit Survey Committee observed that, during 1951-52, there were only two important sources of rural credit, namely the moneylenders and the borrower's relatives who accounted for about 84 percent of the total credit. The government and the co-operatives accounted for 6.4 per cent and commercial banks for just 0.9 per cent[12].

As per the classification usually adopted by the Reserve Bank of India (RBI), there are two sources of credit to farmers in rural India. viz: institutional and non-institutional credit.

The present system of agricultural credit in India is grouped under two headings viz., the institutional and non-institutional. The non-institutional sector consists mainly of the professional and agricultural moneylenders, landlords, traders, commission agent, relatives and friends. The institutional sector comprises of government, co-operatives, commercial banks and regional rural banks. In spite of the significant expansion of the institutional sector, non-institutional agencies continue to play a dominant role in rural credit.

NEED AND IMPORTANCE OF CO-OPERATIVE CREDIT SOCIETY

India is predominantly an agricultural country. About 67 per cent of the people are directly dependent on agriculture. The Indian farmer is not credit worthy because he is poor and he remains poor because he is not credit worthy. The rural co-operative credit agencies as suitable partners are not alien in the context of the development of the economy. The Rural Credit Co-operative Institutions (RCCI), one of the strong arms of Rural Financial Institutions (RFIs) have made significant strides in delivering credit for agriculture and rural development. In other words co-operatives still constitute a dominant segment if not fulfilling the total need of the rural credit market. The co-operative movement in India was a movement against private moneylenders, who were notorious for charging exorbitant rates of interest and oppressive practices, forcing the peasant borrowers into a perpetual state of indebtedness. Initially, experimental credit co-operatives were organised in an effort to break the poverty cycle of small farmers which was accentuated by frequent famines, low productivity and perpetual dependence on the money lenders.

It has been the accepted policy, so for, that the co-operative system is the most suitable institutional agency for this purpose in the conditions of the Indian villages. It is significant to note that agricultural credit provided by the co-operatives is much more significant in quantum than that provided by the other institutional agencies.

PROBLEMS OF STUDY

The adequate and cheap credit is one of the perennial problems of Indian agriculture. Institutional credit for the development of agriculture has made rapid strides over the years[13]. India probably has one of the largest network of co-operative credit institutions supporting agriculture[14]. The important causes for the failure of the movement was the lack of spontaneity of the movement. The co-operative credit system has made considerable progress in the field of agricultural finance but their performance could not come upto the expectation[15]. This is due to the fact that the co-operative credit system suffers from certain serious weaknesses at present. Some of these have been identified as follows.

(a) The continuous existence of weak Central Co-operative Banks has been retarding the progress of the co-operative credit movement in the country. The weak central co-operative banks will not be able to provide the required refinance facilities to the primary agricultural credit societies.

(b) The co-operative banks did not meet all the cash requirements of the farmers.

(c) The co-operative banks were once visualised to take away completely the business of moneylenders as far as short-term needs of the agriculturists were concerned. They were expected to attract deposits just as other commercial banks do and borrow from other members of the money market and make loans and advances to the agriculturists through the village primary societies. But these expectations have not been fulfilled.

(d) Another problem of the Co-operative Society is politicisation. The politicians utilise co-operative organisations as springboards of political power and patronage and hence co-operatives have lost their virtue.

The PACBs provide the wheels to the vehicle of the co-operative banking network. The success of the cooperative movement depends on the inherent efficacy of the primary co-operative credit structure. Therefore any scrutiny of the deficiencies of the co-operative credit network has to be focussed on the working of PACBs. It is against the background of a large rural population, small holdings, low incomes and the listing network of credit that one has to view the problems facing co-operative financing in this country. Co-operative banks face major problems. The production credit requirement potential of the country is always estimated high and the co-operatives are expected by the planners to provide about 50 per cent of the agricultural credit needs of the country. But the co-operative banks are unable to do so because one of the weaknesses of the co-operatives lies in their inability to mobilise deposits in the rural areas. Thus the Co-operative banks in the country lack much in self-sufficiency with a continued dependency on the Central Banks of the country.

One of the major problems of co-operative financing in India has been that of mounting overdues. The higher overdues have affected the overall financial health of the rural credit structure. The amounts defaulted by the borrowers have acted as a serious restraint on the smooth flow of farm credit and constitute a major obstacle in meeting the credit needs which tend to increase year after year. The problem of overdues appears to be more serious with respect of PACBs. The profitability of the co-operatives has been a matter of considerable concern. Unsatisfactory recovery of dues, low margins and non-viable nature of many PACBs are important factors affecting the working of PACBs.

Hence, co-operative banks having less amount of deposits and meagre funds of their own are being crippled with mounting overdues. Thus the inability to expand the volume of credit at the farm level. The present co-operative movement depends upon cheap financing from the apex body, viz. NABARD. Under the sweeping economic reforms which are taking place in the economy whether credit at cheap and discriminative interest will be available to co-operatives in future is a million dollar question. Whether the institution with structural instabilities, mounting overdues, politicisation of movement, inadequate deposit mobilisation can survive on a competitive financial regime based on market efficiency without the government's subsidised interest is a question which lingers in the minds of people.

The gap between the demand for credit and supply of credit by the PACBs is also on the increase and hence farmers have to approach other sources of finance to meet their demands. This is much against the basic objective of the co-operative movement. The entire cooperative movement is at the cross roads. The movement has to be made more viable with least governmental and political intervention. Thus the problem chosen for this study at the PACB levels assume greater importance.

AREA OF THE STUDY

Vellore district has been chosen for the study for the simple reason that it is one of those districts which succeeded in the green revolution and Madras University and Cambridge University projects on Green Revolution had also selected this district. It is again a district dominated by small and marginal farmers and one that experienced the pump-set revolution. There are no big estates in this district. The co-operative credit structure is said to have fairly succeeded in this district.

"North Arcot contributes about 10 percent to the total area and production of rice in Tamil Nadu. North Arcot has the largest number of wells and the highest density of wells. North Arcot is a district of small farmers, the average holding being 1.43 hectares. It has the largest proportion of small cultivators among the districts of Tamil Nadu. Agricultural loans are utilised for all crops, and possibly for non-agricultural purposes also. About 32 percent of the loans are from co-operatives. HYV cultivators obtained 59 percent of their credit from co-operatives whereas other paddy cultivators obtained only 25 percent of their credit from this sources"[16].

Having chosen a representative district to analyse the working of primary agricultural co-operative banks in agricultural credit, wide ranging discussions have been held with the district co-operative authorities and the compiled annual reports of primary agricultural co-operative banks at the district level have been carefully perused.

The villages have been again purposively chosen on the advice of the district co-operative authorities. The present study aims to find out the reasons which have made some PACBs the 'best' and the others as 'bad'. Thus, after looking at the performance of all the PACBs in Vellore district, the district authorities have indicated Thamaraipakkam PACB as good and Kalathure PACB as bad, as per the criteria adopted by them.

PERIOD OF STUDY

The study has covered 10 years from 1989-90 to 1998-99; the analysis of secondary data is for this period. The cut-off year 1989 was selected due to the reason that, Vellore District was bifurcated in 1989 into Vellore district and Thiruvannamalai district. During the study period the New Economic Policy (1991) was also implemented. Therefore, its impact on the working of cooperatives can be studied and hence the post-reform period for the period of study (10 years). The secondary data are available only upto 1998-99 or 2000 and hence the study has been limited to this period. However, wherever recent data are available they have been used for analysis.

SCOPE AND LIMITATION OF THE STUDY

The present study has a limited scope of examining the working of a 'best' and a 'bad' cooperative society and finding the reasons for such a performance. The growth in membership, effective borrowers, quantum of credit extended to farmers, rural savings mobilised have been discussed for both the societies and the difference in the performance of these societies in their functioning have been pointed out. The study makes a brief note an the quantum of overdues and mobilisation of rural savings but has not gone deep into these problems. The study has not attempted to study whether farmers use their credit for the purpose for which it has been sanctioned or any diversion of credit is taking place. The study has not made any attempt to study whether the cooperative credit has made any impact on their income level or even on the pattern of cultivation.

OBJECTIVES OF THE STUDY

With the above said scope and limitations the following objectives have been formulated:

(a) To examine the flow of institutional credit to agricultural sector in the study area.

(b) To study the distribution of the credit (short-term, medium term and long-term) and size of farms.

(c) To study the savings mobilisation among the farmers by the PACBs.

(d) To study the position of overdues at various levels, their causes and remedial measures.

(e) To suggest measures to improve the co-operative credit movement.

HYPOTHESES

Based on the above said objectives following hypotheses have been formulated:

(a) Size of credit is directly related to size of farm.

(b) The co-operative credit has not grown to the needs of farmers.

(c) The overdues are inversely related to size of land holdings.

(d) The co-operative movement has made little progress in mobilising rural savings.

(e) The co-operatives have diversified the rural employment activities.

SOURCES OF DATA

The materials for the study have been collected from various sources. The data on the state and the district economy were collected from policy reports of Government of Tamil Nadu, Tamil Nadu: An Economic Appraisal, published by the Evaluation and Applied Research Department and the Statistical Hand Book of Tamil Nadu published by the Directorate of Economics and Statistics. Data on co-operatives have been collected from the office of the Tamil Nadu State Co-operative Bank (TNSC Bank), Registrar of Co-operative Societies, Chennai, Vellore District Central Co-operative Bank, the Agriculture Department and from the two PACBs. The gazetteer of North Arcot and the District manuals were also used to supplement the data. Information on the villages of the two PACBs, has been collected from the offices of the District Collectorate, District statistical department and from the offices of Arcot and Arakkonam Panchayat Unions. In addition to these sources, various reports on co-operative credit, books and journals were referred.

Primary data has been collected from villages of Thamaraipakkam PACB and Kalathure PACB, by using a structured questionnaire.

In addition, a comprehensive survey of literature in the relevant fields has been made to formulate the framework. The researcher has made an intensive use of the Libraries of University of Madras, Presidency College, State Planning Commission, RBI Staff Training College, Chennai, Madras Institute of Development Studies, Chennai, Connemara Public Library, Chennai, and Natesan Institute of Cooperative Training Centre, Chennai. Apart from these, the researcher has also visited research Institutes and met experts in the field to interact with the faculty members.

METHODOLOGY

To analyse the benefits of co-operative credit among the various categories of members of PACBs in Vellore District, the analytical framework used both primary and secondary data. There are 185 primary agricultural co-operative societies functioning under the guidance and control of Vellore District Central Co-operative Bank. Two PACBs have been selected for the study and have been identified after a thorough discussion with official's who have indicated the 'Best' and the 'Bad'. The best PACB is Thamaraipakkam PACB and the 'bad' is Kalathure PACB. The identification of PACBs are based on their performance. The official's opinion has been strengthened by primary and secondary data. This study covers the period from 1989-90 to 1998-99, a period of 10 years, however, all secondary data has been updated and cover 13 years wherever possible. The officials have identified on the basis of some criteria, such as overdues, deposit mobilisation, issue of loan amounts, number of loans borrowed or effective members, etc. The basic characteristics of these villages and PACBs are discussed in the concerned chapters.

Having chosen a representative district to analysis the working of primary agricultural co-operative banks in agricultural credit, wide ranging discussions have been held with the district co-operative authorities and the compiled various annual reports of primary agricultural co-operative banks on district as well as village level have been carefully analysed.

A sample of 371 members were selected from the selected primary agricultural co-operative banks by adopting systematic random sampling. A 10 percent of total members in the PACBs (total membership in Thamaraipakkam is 1530 and Kalathure PACB is 2181 during 1998-1999) have been selected as sample in consonance with the objectives. The total members were classified on the basis of size of land holdings and every 10th member has been selected. Thus 10% of every category of farmers are represented in the sample. A proportionate sampling method was adopted to select various categories of land holding farmers in the same proportion in which the universe of sample is constituted. A sample of 371 was selected for interview on the basis of systematic random sampling method with equal probability for every one in the universe. The analysis of primary data collected through interview schedule is supplemented by analysis of secondary data, field visits and discussions with farmers, officials etc.

Since most of the respondents happen to be illiterate, the data required were collected through interview schedules using a set of pre-tested questionnaire and standardised techniques of recording. Standardisation ensured that all respondents replied to the same question. Nevertheless, mostly close-ended questions were used. The sample households were personally interviewed by the researcher over a period of six months during 2000-2001, which yielded a close insight into their credit needs and their agricultural and allied operations.

The collected secondary and primary data have been computerised and using relevant statistical tools, the data has been analysed.

For analysing the quantum of credit borrowed, repayment behaviour, overdues, credit gaps etc, primary data were obtained by conducting a field survey. The data thus collected from the beneficiaries were suitably tabulated, analysed and interpreted by using statistical tools like percentages, mean, 't' test, regression analysis, multiple regression technique, chi-square test, analysis of variance, correlation co-efficient, etc.

CHAPTER SCHEME

The study is divided into eight chapters. The first chapter deals with the importance of the study, nature of the research problem, objectives and hypotheses. It also gives a detailed description of the methodology, area of study and limitations of the research.

Second chapter deals with review of literature. This consists of two parts: (i) empirical studies in this area, and (ii) various committees pertaining to this study.

Third chapter gives a brief discussion on the co-operative movement the world-over and at the India level.

Fourth chapter is devoted to the general features of Vellore district and two PACBs. It gives general theme of agricultural growth. Pattern of cultivation, occupational distribution, land use pattern, institutional role in the district level and at the level of two sampled PACBs villages.

Fifth chapter captures the characteristic features of secondary data from Vellore district central co-operative banks and selected two PACBs.

Sixth chapter deals with the existing facility and the problems faced by the society members. This analysis is based on the primary data collected by the researcher.

Seventh chapter presents an analyses of working of sample PACBs and the test of formulated hypotheses.

Eighth chapter provides summary and conclusions based on major findings of the study. Some suggestions to be considered for better functioning of the credit society are also presented in this chapter.

REFERENCES

1. Cohen, R.L., *The Economics of Agriculture*, Cambridge University Press, London, 1965, p.1.
2. Bhatia, B.S., VERMA.H.L., and Mahesh C. Garg, *Encyclopaedia of Co-operative Management, Co-operative Banking: Levers of Rural Economy*, Vol.3, Deep and Deep Publications, New Delhi, 1994, p.237.
3. Sivasubramanian, A.P., *The practice of Co-operative Banking*, Arul Selvi Publications, Tirunelveli, 1986, p.5.
4. The Hindu, Survey of Indian Industry - 2003, p.12.
5. Statistical Outline of India, 1999-2000. TATA service Limited, Department of Economics and Statistics, Mumbai, 1999.
6. Raj Kishor Pany, *Institutional Credit for Agriculture in India*, Ashish Publishing House, New Delhi, 1985. p.
7. Ghosal, S.N., *Agricultural Financing in India*, Asia Publishing House, Bombay, 1929, p.27.
8. Krishna Kumari, A.B., *Co-operative Banking for Agricultural Development*, Deep and Deep Publications, New Delhi, 1998, p.51.
9. Ciarence Zuvekas, *Economic Development - An Introduction*, The Macmillan Press Ltd., London, 1979, p.206.
10. Ranga Reddy, A, *Agricultural Development Rural Credit and Problems of its Recovery*, Mittal Publications, New Delhi, 1990, p.162,168.
11. Kamta Prasad, "Financial Intermediaries and Rural Development", *The Indian Economic Journal*, Special conference number for 1978, Part-II, Vol.26, No.2, October-December, 1978, p.287.
12. Bansil, P.C., *Agricultural Problems of India*, Oxford and IBH Publishing Co, New Delhi, 1981, p.224-228.
13. Vasant Desai, *Rural Economics*, Himalaya Publishing House, Bombay, 1983, p.400.
14. OP.CIT. Ranga Reddy.A, p.162, 105.
15. Mudgal, M.K., "Co-operatives at Cross-Roads," *Kurukshetra*, Vol.46, No.8, May-1998, p.15.

16. OP.CIT. BHATIA B.S., VERMA.H.L., and MAHESH C. GARG, p.137,246.

17. FARMER. B.H., *Green Revolution? Technology and change in Rice-Growing Areas of Tamil Nadu and Srilanka,* The Macmilla Press Ltd., London, 1980, p.92,94.

Chapter - 2

REVIEW OF LITERATURE

INTRODUCTION

In this chapter an attempt is made to give a brief review of literature. The literature available on this title is exhaustive and the committees and commissions involved in this field are numerous. Hence, an attempt is made here to trace the available literature as brief as possible. The survey has been made in two different sections : one for individual studies and another for various committees. For a better understanding the studies have been reviewed under seven different headings.

AGRICULTURE FINANCE

BHALERAO *et al.*, (1968)[1] studied the problem of regional imbalance in the process of growth of co-operative movement. He had recommended for greater emphasis on accelerated development of the co-operatives among under-developed states with more finance, trained manpower, efficient long-term credit management and more stringent measures to minimise the mounting overdue problem which inhibits the recycling of the funds for increased credit facilities from the Land Development Banks.

TYAGI (1968)[2] had revealed that the formula of co-operation, namely "one society to one village and one village to one society" has failed in India. The encouragement of savings, and specially in the rural areas, one of the foremost aims of the co-operative credit movement, has perhaps been the most intractable problem. Savings must come from a habit of thrift but probably it was not generally recognised that thrift is a business-like quality of which the first implication is the existence of business.

MITRA and SAIN (1982)[3] observed that the growth rates of all the variables relating to the supply of finance by the co-operatives in the area under study (Burdwan District) were not uniform, and did not equally stand the test of significance. The growth rates of total advances in money terms and of number of borrowers had been found quite high during the period under study (1982). The progress of total advances in real terms was also quite satisfactory.

A study by CH.BHAGAWAN RAO and KANAKAIAH (1983)[4] concluded that, (a) the total membership, the dominant part is from the membership of marginal and small farmers. (b) The number of agricultural labourers is not as impressive as members from small farmers.

MATHUR (1983)[5] has pointed out that the development of institutional credit is a basic condition for agricultural progress. The history of agricultural development in all advanced countries shows that an integrated system of institutional credit laid the foundation of agricultural prosperity. The objective of the institutional credit is to make a breakthrough in the vicious circle of poverty, rack-renting, usury and debt and to stimulate the farmer to boost agricultural productivity.

HANUMANTHAIAH *et al.*, (1986)[6] attempted to find how the financing agency is functioning in Prakasam District of (Andra Pradesh), which provided 99 per cent of the advances to the priority sector alone. Based on their analysis, it is inferred that the farmers could not realise benefits in proportion to their investment. The cost-benefit ratio indicates that the farmer's realisation per rupee basis declined alarmingly.

BAIDYANATH MISRA (1997)[7] has concluded that with liberalisation, a slow wind of change is now taking place in the market place and the competition therein. Co-operative Banks now face competition not only from the commercial banks but also from the international banks. A potential threat looms largely not only from the indigenous companies but from the foreign companies. With the concessional finance becoming more difficult to get from the govt., the emphasis has been shifting to mobilisation of more resources and its availability for investments.

KRISHNA KUMARI (1998)[8] concluded in her study that the small farmer and marginal farmer face the problem of restricted credit facility as compared to the large farmer whether it be in the informal money market dominated by merchants, professional money-lender and landlord money-lander (or) the institutional credit supplied by commercial banks and co-operative institutions. The ratio of informal credit and institutional credit provided to farmers varies from country to country. Unfortunately, in evolving credit programmes, small farmers are at great disadvantage as compared to the large farmers who are appropriating most of the credit facilities. The operation of the credit system reveals a bias in favour of large farmers for a few important reasons.

MUDGAL (1998)[9] concluded that the co-operatives had played a very important role in our agriculture system. Credit co-operatives had a considerable share in the success of Green Revolution. Similarly, in bringing about the White Revolution, co-operative's role is well known to the world. To imagine Indian agriculture without co-operatives is really difficult.

RURAL CREDIT POLICY

ATCHI REDDY and PRABHAKAR (1970)[10] observed that the per acre credit increased along with the size of the holding due to a host of factors. The increasing role of the organised credit came in handy and saved many farmers from falling into the clutches of the moneylenders. The expansion of the organised credit market largely helped the farmers to increase their market surplus and thus, to achieve a degree of commercial farming, which is at present all set for take-off into a fast growing agriculture.

MAMORIA and SAKSENA (1977)[11] suggest that the reduction of overdues, mobilisation of deposits, revitalisation of the dormant societies, amalgamation of the weaker ones into viable units, introduction of crop loan system, effective linking of credit with marketing and improving the overall efficiency of co-operatives by toning up their management and better supervision of the use of loans at all levels are the most important aspects of future co-operative policy. If these are accomplished, co-operative societies may cease to be a mere group of borrowers as at present and emerge as effective banking agencies and serve the cause of agricultural production better than at present.

RAJKISHORE PANDA's (1979)[12] study pointed out that the large farmers received a greater percentage of credit than the marginal and small farmers. The additional percentage increase in loan amount is found to be less than the additional percentage increase in loanees, implying that inadequate credit financing of

PACB per loanee in the year 1982-83 as against 1979-80. He concluded that the co-operative credit distribution was not only uneven between different size classes of holding but also the growth in amount of credit per borrower member in all categories of farm households during the period of study was very much inadequate.

RATNAPARKHE and PAGIRE (1980)[13] have reviewed the requirement of credit for various crops and the present credit limit fixed by the different institutions. The credit gap was smaller for bajra and groundnut. A little enhancement in the scale of finance for these two crops will adequately bridge the gab. However, in the case of sugarcane, a wide gap existed between the cost of cultivation and the scale of finance. There is a need to enhance the scale of finance for sugarcane. No consistency was observed between the scale of finance and the actual cost of cultivation.

BEOHAR and KHARE (1982)[14] have concluded that farmers who obtained credit for the purchase of pump-sets repaid the loan in the beginning but those who received loan for the purchase of fertiliser and digging of wells did not repay the loan in proper time. A large proportion of loans advanced for the purchase of bullocks and for land improvement remained overdue as compared to other purposes. The farmers who adopted a large number of packages of production practices, irrespective of the purpose of credit, had higher repaying capacity per farm as compared to others. The farmers who obtained loan for pump-sets had the maximum repaying capacity as compared to other purposes.

A study by SURYAWANSHI (1982)[15] revealed the fact that the use of high cost inputs, resulted in increased demand for credit. The high cost was associated with high intensity of cropping and more cash crops. Hence, the financing institutions should provide more credit to the agricultural sector. The working expenditure on the cost of cultivation of crops must be taken into consideration in deciding the loan policy.

DINESH K. MAROTHIA (1985)[16] based on the farmers opinion, has concluded that high input prices, timely and adequate availability of agricultural inputs and credit, desired type of input were the most important problems in the input supply system of PACBs. But PACBs managers have stressed on the following management problems: unmanageable working zone, inadequate staff, inadequate input storage facilities, improper utilisation of loan, conversion of productive loan into mass exemption of loan, influence of socially and economically powerful farmers and poor co-ordination with other departments.

SRINIVASAN (1985)[17] observed that 74 per cent of the credit needs of the farmers were met by institutional credit agencies. Their enquiry indicated that the farmers felt that procedural difficulties, the rigidity (or) delayed disbursement of credit, the prohibitive cost of acquiring it and its inadequacy were the major impediments in getting long-term finance from the institutional agencies.

GANGADHARAN (1986)[18] pointed out the high cost of borrowing, especially, for small borrowers due to enhanced premium on risk and administrative cost. Further, he observed that the small farmers were not responsive to lower interest rate and such policy of providing subsidised interest rate may vitiate the dynamism in the capital market.

AMRIK SINGH SAINI *et al.*, (1988)[19] reveal that the integration of crop farming with high-yielding milch animals markedly increased the income and employment of marginal and small farmers. Dairying, being a self-income generating enterprise, reduced the short-term credit requirements by supplying regular income to the farmer but increased the medium-term credit requirements for the purchase of the high-yielding milch animals. The financial institutions should, therefore, come forward, unhesitatingly to provide medium-term loans on easy terms to these farmers for the integration of their arable farming.

MANIDORASWAMY NAIDU and SURENDRABABU (1988)[20] observed that the institutions in the developing countries follow the traditional credit need approach while lending. This approach appeared to be defective and misleading. Due to this, loans advanced to the target groups were misused.

VAIDYA *et al.*, (1988)[21] forcibly argued for the need for the formulation of liberal policies for efficient distribution.

ALAMELUMANGAI (1995)[22] pointed out that the recent credit policy of the RBI has deregulated the interest rate for co-operative banks. Now, these banks are permitted to use their discretion in stipulating the rate of interest both for deposits and credit. As the gate is opened only for co-operative banks, it is the opportune time for them to make use of this facility to the optimum level to improve their performance at present and compete successfully with other financial institutions in the near future.

STRUCTURE AND GROWTH OF CO-OPERATIVE CREDIT SOCIETY

SATISH VERMA and VIJAY CHOPRA (1968)[23] in their study, suggested that more bank branches should be opened in this sector of the economy so as to meet the unfulfilled demand for banks credit. The study also pointed out that as the demand for bank credit was inversely related to the national income originating in this sector, therefore, more funds can be made available to the small and marginal farmers. Further more, the rate of interest mechanism could be used to encourage the flow of credit to the small and marginal farmers. The demand for bank credit was positively related with the adoption of new technology and the cost of production in this sector of the economy.

PANDEY and ASHOK KUMAR's (1970)[24] study revealed that per hectare credit availability on small and marginal farms was higher than on farms above 2 hectares in almost all states. However, on per borrower basis, the position was reverse. The study, further, revealed that the disparity in distribution of credit on per hectares basis has increased. The important factors affecting the flow of credit in the states are fertiliser consumption, irrigated area, application of high-yielding varieties and area under cash crops.

PANDEY and GUGLANI's (1970)[25] results revealed that in the advancement of agricultural loans by the various financial institutions, a positive trend was noted, for both the tribal and non-tribal farmers during the period under study. Linear growth rate of agricultural loans was found to be 23.56 and 22.92 per cent in the case of tribal and non-tribal farmers respectively. The results showed that the outstandings of agricultural loans advanced by the financial institutions were mounting high year after year in the case of non-tribal farmers. An increasing trend was observed in institutional lending to the tribal farmers since the inception of 'Multi-Agency Approach' in rural credit.

RAMACHANDRA BHAT and ARUNA RAO (1982)[26] observed that the evolution of the co-operative credit policy in India indicated that a definite policy of integrated co-operative credit emerged by the end of the first five year plan. Co-operatives are the most important agencies supplying largest amount of institutional finance to agriculture. According to them, the positive rate of growth of overdues implied a negative or slow performance on the part of primary credit societies.

SARKER *et al.*, (1986)[27] pointed out that the institutional agencies in rural areas have not shouldered the full responsibility of providing credit facilities for the sake of rural as well as national development. His suggestions were that commercial banks, co-operatives, government institutions should cover more number of small and marginal farmers by relaxing the existing norms and terms for procuring credit. The volume of credit should be sanctioned not only on the basis of cost of cultivation (or) size of farming; but consumptional needs of the farmers are to be considered.

SHARMA and TEWARI's (1986)[28] study pointed out that when the large farmer got more credit under the area-based allocation, it generated more wage employment opportunities for the marginal farmers to whom wage income forms a substantial part of income. Hence, this allocation criterion might cause the largest reduction in the inequality of income distribution.

GOYAL and PANDEY (1988)[29] dealt with the analysis of the data presented on the growth trend of certain parameters of primary agricultural credit co-operative societies. They concluded that there was an upward trend of all the parameters. But the highest compound growth rate was recorded for overdues which was not a healthy sign of co-operatives. This reflected upon the poor rate of recovery.

GANESH P. GUPTA and SUDHA RANI SAHU (2000)[30] pointed out that, the primary co-operative societies available for rural development task, have to re establish the faith on the co-operative system and then the democratic set-up will get strengthened. The involvement of co-operative societies and their capacity to provide effective representation and services for farmers / non-farmers / rural artisans, alike, especially small producer, is a key factor in steering loyalty of the people. Co-operative societies are required to enlarge the scale of business operation. Increased business operation will fulfil the expectation of the people, which will in turn, strengthen the loyalty in co-operatives; for increased business operation will yield more economic gains and prosperity.

ROLE OF CO-OPERATIVE SOCIETY

MAMORIA and SAKSENA (1977)[31] revealed that the primary co-operative credit society is intended to promote the economic interests of its members in accordance with the co-operative principles and this aim is achieved by activities in different directions such as promoting savings among members, providing loans to them, supplying them with agricultural requisites and domestic requirements and arranging for the marketing of their agricultural produce. For this purpose the single-purpose societies have yielded place to multi-purpose societies.

MATHUR's (1980)[32] study pointed out that the co-operative organisations have been recognised as the best institutions for providing rural credit to the farmer because they satisfy all the important criteria of sound agricultural credit. It has been the accepted policy so far that the co-operative system is the most suitable institutional agency for this purpose in the conditions of the Indian village. It is significant to note that the agricultural credit provided by the co-operatives is much more significant in quantum than that provided by the other institutional agencies.

AGRICULTURAL CREDIT BOARD, RBI (1981)[33] has instructed that there is an increased emphasis on re-organised primary agricultural credit societies undertaking multi-purpose activities like meeting the consumption credit needs of the weaker sections of the rural community, distribution of agricultural inputs, marketing of agricultural produce, distribution of consumer articles in rural areas, etc., for which the society requires additional resources.

A study by SANKARAIAH and MUNIDORAISWAMY NAIDU (1983)[34] revealed that the role of co-operatives in meeting the increasing financial requirements of small farmers in the wake of higher costs of cultivation under new technology. The security oriented credit policy of the co-operatives and dominance of large farmers in the governing body of co-operatives are the main hurdles in getting the loans.

RAYUDU (1984)[35] observed that today co-operative banks are solving divergent problems of small farmers, marginal farmers and agricultural labourers. By identifying the small, marginal and agricultural labourers the co-operatives are giving more short-term whereas commercial banks are giving more medium and long-term credit.

SATYANARAYANA *et al.*, (1986)[36] observed that to some extent, the subsidy enhanced their incremental income and it was proved by their early repayment mostly by landless labourers and marginal farmers. According to them, it seemed necessary to initiate certain measures, in providing subsidy to identified beneficiaries by the farmers service society.

VISWANADHA RAO and MOHANDOSS (1986)[37] concluded that despite their short-comings and weaknesses, the co-operatives had done fairly well in the field of rural credit. However, certain internal and external problems were hampering their smooth progress. The internal problems related to high overdues, non-borrowing membership, non-viability, dormancy, lack of professional management etc. External problems arose out of lack of adequate infrastructure, instability of agriculture, scattered and uneconomic holdings, etc. The co-operatives possess the organisational advantage of being located in the villages by virtue of which they can serve the rural masses effectively.

SINGHAL'S (1988)[38] study revealed that, in pursuance of the policies of the state government, the society has made credit arrangement for people of different walks for their various economic activities. There is no class of society to whom the money is not available.

DANDEKAR (1994)[39] pointed out that the entire co-operative credit structure has been built on a weak base, namely, the primary credit societies at the village level, without realizing, (or) at least openly admitting, that the primary societies are weak because their lending business is essentially non-viable. Hence, little was done except to exhort that the primary societies be revitalized. Now, if we wish to take seriously the suggestion of the committee on financial system (1991) that the concessional credit to the non-viable sector should be gradually phased out, it will amount to gradually winding up co-operative credit institutions at all levels as privileged institutions and bring such of them as will survive at par with the commercial banks.

MAMORIA (1995)[40] has concluded that the co-operative credit societies were started to carry on the business of rural banking and replace the moneylenders, yet they have not made any appreciable progress.

GOPAL LAL JAIN (1997)[41] reveal that, primary agricultural societies operates at the village levels and maintains a direct contact with the farmers. The PACB is the kingpin of the agricultural co-operative movement in India. The government made attempts to nurture the co-operative movement to institutionalise efforts to relieve farmers from the traditional burden of debt and to promote thrift. Gradually, they assumed a more positive role as compared to their earlier characteristic of organisation for defence against exploitation by money-lenders. There was not only a steady quantitative expansion in numbers but also a growing diversity in the functions assumed by co-operative societies.

CHANDRA PRASAD and VISWESWARA RAO (1998)[42] concluded that the establishment and growth of co-operatives is regarded as one of the important instruments for economic, social and cultural development as well as human advancement in developing countries. Co-operative credit institutions and banks have emerged as a ray of hope and vehicles of rural finance for economic development.

VERMA (1999)[43] concluded in his study that co-operatives are a vital instruments for economic and rural development. It is the best instrument to promote social justice and prevention of exploitation at the grassroot level. Weaker sections have always been the thrust area of the co-operatives. Really, the co-operative is the only way to improve the rural poverty in productive ways.

UTILISATION OF CO-OPERATIVE CREDIT

BALISHTER and CHAUHAN (1982)[44] estimated the extent to which co-operative credit was utilized for productive purposes on various categories of farms. The study revealed that per farm availability of crop loan from co-operative society under study increased with the increase in farm size while a reverse trend was observed for per hectare. The crop loans were advanced both in cash and in kind. But the proportion of loans advanced in kind was much higher than the proportion of loans advanced in cash. This is mainly, to check the diversion of credit to unproductive uses.

CHAUHAN *et al.* (1982)[45] estimated that the small and medium farmers availed both production credit and investment credit from banks while large farmers availed only production credit. The majority of farmers availed loans both from co-operative and commercial banks in all the categories. In case of non-institutional credit, the small and medium farmers availed credit both for productive and unproductive purposes. But the extent was higher for the latter while large farmers availed loans for the productive purpose.

RADHAKRISHNAN and MUKUNDAN (1984)[46] found an inverse relationship between the amount of loan per hectare and the size of holdings. Small holders obtained relatively more amount of credit than that of large holders.

GUPTA and VERMA (1986)[47] concluded that the utilization of co operative finance for tractors, tube-wells, pump-sets and wells enabled the farmers not only for repaying the credit with interest from the additional returns but also their financial status improved.

RAJPUT (1986)[48] concluded that in case of non-institutional finance, the minimum proportion came on large farms being more resourceful and their needs were fulfilled by institutional agencies. It was minimum on marginal farms. The overall average loan availability for production and investment came to about 26 per cent and 74 per cent respectively. The investment credit was taken primarily for the development of irrigation infrastructure.

VAIKUNTHE (1988)[49] studied the utilisation of co-operative credit by all the size groups in the irrigated area. According to his study a small amount was being misutilised in non-agricultural purposes.

KALIYAPERUMAL and LOGANATHAN (1989)[50] pointed out that the farmers provided with the required amount of credit in full, without any gap, their production would have been better and their financial standing would have been improved very much.

BAIDYANATH MISRA (1997)[51] reveals that the credit to the needy agriculturist is the major function of the co-operative societies. Co-operative credit societies have been production oriented and issued loans mostly for purpose of seasonal agricultural operations, through PACS.

ON IMPACT ON PRODUCTION AND INCOME

KALYAN BHATTACHARYA (1978)[52] concluded that the failure of co-operative societies becomes all the more glaring in view of an unmistakable rise in the importance of credit in development programmes of agriculture; on the other hand the importance of land and labour has declined considerably in recent years.

ADINARAYANA and LAKSHMINARAYANA's (1983)[53] study showed that the quantum of loan in the case of small farms under co-operatives was not adequate while in all other farm categories it was just sufficient.

SAMIUDDIN and MAHFOOZUR RAHMAN (1983)[54] suggested that, the co-operative farming is to be encouraged as rapidly as possible so that in ten years or so a substantial proportion of agricultural lands would be so cultivated. The promotion of village and small industries through co-operative organisations hopes to increase employment, avoid technological unemployment wherever possible and to provide the basis for the structure of an essentially decentralised society.

SRIVASTAVA and LAL (1983)[55] observed that crop loans and borrowings for livestock have not made any significant impact on progressive farms. On less progressive farms, the livestock elasticity was positive and significant at ten per cent level of probability, indicating that farm credit for livestock significantly contributed to this category of farm.

SINGH *et al.* (1985)[56] conducted the per farm business analysis which revealed that the value of input, output, family labour income, farm business income and input-output ratio were significantly higher on the beneficiary (this study is based on a random sample of 100 cultivators, 50 beneficiaries and 50 non-beneficiaries) farms in all the size-groups as compared to the non-beneficiary farms.

LAL and LAVANIA (1986)[57] observed that the farmers owning large holdings are, generally, the greatest beneficiaries of the co-operative credit. Major part of the credit is utilised for the productive purpose in agriculture.

HEMCHAND JAIN (1988)[58] studied the impact of farm credit on the income of tribal farmers through increase in production and employment in selected survey area.

SINGHA and SHARMA (1988)[59] pointed out that there was scope for increasing farm income through optimal reorganisation of available resources and adequate provision of production credit for farmers.

BHATIA *et al.* (1994)[60] pointed out that the co-operative banks have played a vital role in the development Indian Economy. Green Revolution in states like Punjab would not have been possible, if co-operative banks had not provided timely and adequate credit to the farmers. However, there is a large/potential to expand and improve co-operative banking system in India.

GULAB SINGH AZAD (1997)[61] pointed out that the contribution of co-operatives in raising agricultural productivity has been on account of increased infrastructure facilities and supportive service like agricultural credit, supply of agricultural inputs like fertilisers, improved seeds, insecticides, irrigation facilities etc. He concluded in his study, that co-operatives have been successful to a large extent in this and today their share in agricultural lending stands at 44 per cent approximately.

REPAYMENT CAPACITY

DUHAN and SINGH (1977)[62] observed that the reduction of imbalance in the distribution of co-operative credit among different categories of households. His major finding was that the small farmers registered a higher percentage of default as compared to the marginal farmers and landless labourers.

BHOSAL and BANGAT (1981)[63] in their study suggested the important reasons for non-repayment reported by the farmers were their low income, non-remunerative prices for agricultural produce and crop failure due to natural calamities. The results of the multiple linear regression analysis revealed that the amount of overdues was related with the net income from crop production, amount borrowed, amount used for productive purposes and expenditure on family consumption in all the groups of holdings. This calls for appropriate policy with regard to supervision of credit used at the farm level.

GANGWAR and KUSUM AGGARWAL (1981)[64] observed that the main causes of default in recovery of loans : are (a) excessive political interference, (b) supply of sub-standing farm inputs, (c) wilful default (d) repaying capacity fell short of loan amounts due, (e) fictitious loan, (f) lack of adequate income to repay the tractor loans and loans to non-viable projects.

VASANT DESAI (1983)[65] concluded that the problems of recovery engender far-reaching and long-term consequences. A considerable portion of the overdues of co-operative credit institutions has been due to wilful default in the repayment by the borrowers.

DATTA (1985)[66] assessed that the default rate would also be uneven. At least, the extent of high overdues can partially be reduced if an element of cushion is provided in devising the feasible repayment schedule.

JULKA *et al*'s (1988)[67] study revealed that the repayments of loans were more satisfactory in the region with transformed agriculture than those in the region with traditional agriculture. Prosperity in the region with transformed agriculture has encouraged quite a number of farmers to incur debt for the purchase of unproductive assets.

RAGHU RAM and BHAVAN DEVI (1988)[68] concluded that the prompt repayment tradition is there for several years in the village where there is a strong village leadership.

KHUSRO (1989)[69] pointed out that the incidence of overdues in the agricultural credit system has been increasing over the years and has been steadily eroding the financial soundness of the system. The increasing incidence of overdues being faced by all the agricultural credit agencies, is very closely connected with the complex nature of agricultural economy and unless the basic issues involved in agricultural sector are tackled effectively, it will be difficult to find a lasting solution to the problem of overdues.

RANGA REDDY (1990)[70], concluded that the overdues position of small and marginal farmers was almost constant. The percentage share of small and marginal farmers in the short and medium-term loans by co-operatives was greater than their share in overdues. It implies that small and marginal farmers are prompt in repaying loans, when compared to other constituents of agriculture. The co-operatives have failed to meet the investment, production and consumption credit needs of small farmers. The small farmers are prompt in repaying their loans compared to the other categories of farmers.

DUTTA (1991)[71] pointed out that the delay in issuing loan is a major problem faced by the members. Consequently the loan cannot be used productively; it is utilised for consumption purposes and this makes repayment of the loan more difficult.

RADHAKRISHNAN (1994)[72] concluded that only 14 members out of 47 loanees are defaulters. The defaulting amount comes around Rs.54,820, the year 1990-91. The remaining 33 members are prompt payers. He suggested that the crop insurance scheme should be widely published and made applicable to all part of the country. To make the crop insurance attractive, PACBs should come forward to pay a part of the premium of insurance policy.

ANANDHARAJAKUMAR and AROCKIARAJ (1995)[73] mentioned in their study that the PACBs are facing a serious situation on account of mounting overdues. The crucial problem of mounting overdues in credit structure has been inhibiting credit expansion to the desired extent. Increasing overdues have eroded the capacity of PACBs to extend further credit for want of recycling of funds and ineligibility of the institutional agencies to borrow additional funds from the higher financing agencies.

HAND BOOK OF INDIAN AGRICULTURE (1995)[74] reveals that the agricultural credit sector in our country is suffering from stunted growth and poor recoveries. Despite the increase in overall agricultural credit, there is a serious problem of overdues which has been inhibiting credit expansion and economic is viability of the lending institutions, especially the co-operatives and the RRBs.

RANGARAJAN (1997)[75] mentioned in his study that during 1980s the rate of growth of overdues was more than the rate of growth of loans. As per the data recovery of agricultural dues to demand stood at 59 per cent in the case of Primary Agricultural Co-operative Banks, 61 per cent for LDBs, 58 per cent for RRBs and around 61.8 per cent for Public sector banks. The recovery under poverty alleviation programme has been much lower. Their problems have been further compounded by a pervasive control of financial and administrative nature by the state governments over the years and progressive officialisation has caused considerable damage to the co-operative character of the system.

TAMILMANI and MANIVEL (1997)[76] have pointed out that the timely recovery of loan amount from the borrower-members assumes paramount importance for the sound and efficient functioning of Primary Agricultural Co-operative Banks (PACBs). Financial viability and operational stability of PACBs are highly attributed to the timely recovery from borrowers.

KRISHNA KUMARI (1998)[77] pointed out that the accumulation of overdues has become a chronic problem for almost all co-operative institutions in several countries success.

RBI (1998-99)[78] has suggested that the mounting overdues of the co-operative institutions have severally restricted their capacity of recycling funds for promoting faster agricultural growth. High overdues also make these institutions ineligible for availing of refinance facility from NABARD.

VARIOUS COMMITTEES

The Co-operative banking system in India has a widespread and elaborate organisational structure. On the basis of a consensus opinion that short-term and long-term credit cannot be supplied by the same organisation, the co-operative banking system was developed with two arms. The short-term structure extending production credit has PACBs at the grass-root level, CCBs at the intermediate level and SCBs at the apex level. A number of committees reviewing the performance of the rural credit delivery system recognised the problems associated with this structure and recommended single window system for providing agriculture credit. The Government of India and Reserve Bank of India were to play an active role in the supply of rural credit since the launching of first five-year plan in India in 1951. Many expert committees were appointed from time to time to improve the flow of credit from institutional sources.

Saraiya Committee (1945)

In 1945, the Government of india appointed the Co-operative planning committee, generally known as the Saraiya Committee. This committee made the following main recommendations.

(a) Co-operation should embrace the entire life and all the economic activities of the farmers for raising their overall standard of living.

(b) Within the next 10 years at least 50 percent of the villages and 30 per cent of the rural population in the country should be covered.

(c) The minimum membership of a village credit society should be fixed at 50.

(d) Sale and purchase or marketing societies should be organised to handle the surplus produce of the members[79]

All these recommendations were accepted by all the Registrars of the country at a conference.

Damle Committee Report (1958-59)

On a Resolution of the National Development council, the Government of India Constituted a working group on Co-operative policy under the chairmanship of Mr.K.R.Damle (1958). It recommended:

(a) A primary society should serve one revenue village only including its hamlets, if any.

(b) Storage was bound to assume additional importance as a result of state trading and accordingly village societies would naturally require their own godowns. While it might not be possible to provide a godown for each village society, it can be provided for a group of about 10 villages.[80]

Gorwala Committee (1951)

The committee appointed by the RBI in 1951 under the chairmanship of Shri.Gorwala to plan, organise and supervise a survey regarding facilities available in rural areas for providing agricultural credit to the agriculturists and to make necessary recommendation. The committee submitted its report in 1954. The survey revealed that the shares of institutional and non-institutional sources of rural credit were 7.3 per cent and 92.7 per cent respectively. The committee recommended as following manner:

(i) Production oriented short-term credit called "The crop loan system", based on crop outlay and not on the basis of ownership of assets as done in the post.

(ii) Creation of new institutions like National Co-operative Development and Warehousing Board, the All-India Warehousing Corporation and State Warehousing companies in order to promote storage, processing and marketing facilities.

(iii) Amalgamation of the Imperial Bank of India and certain state-associated banks into the State Bank of India to provide vast remittance facilities for co-operative and other banks and to promote co-ordination between the rural and co-operative credit, and the banking system; and

(iv) Full co-ordination between credit and other economic activities, especially marketing and processing.[81]

Mehta Committee (1960)

In 1960, the committee on co-operative credit (under the chairmanship of Shri V.L.Mehta) recommended that terms of grant of loans by the RBI to co operative banks should be liberalised.

Accordingly, the central co-operative banks became eligible for loans from the RBI through their apex banks. To enable co-operative banks to meet the increasing working capital requirements for production of crops, the bank has enhanced eligibility limits for sanction of credit for banks on the basis of the audit classification 'A'. The rate of interest charged is 2 per cent below the bank rate. It is concessioned if the finances are made for seasonal agricultural operations against government and other trustee securities but it would be higher if it is obtained for general banking purposes even against the same security.

The committee also been recommended as following suggestion:

(a) Separate staff should be provided for liquidation work and for effecting recoveries. The strength of such staff would depend upon the quantum of work.

(b) The responsibility for the supervision of primary credit societies should be given to the central co-operative banks.

(c) The committee recommended that suitable training may be given to the departmental as well as institutional staff and re-orientation courses may be arranged after every five to seven years.[82]

Patel Committee (1961)

The government of India appointed this committee in July 1961 to examine the question of organisational, procedural and administrative difficulties associated with routing taccavi loan other facilities of the government through co-operatives. The committee had submitted its report in 1963. The committee recommended that all loans for normal production should be arranged through the co-operative institutional agency and that government would have provide finance directly to the farmers only in certain cases of high financial risks. The government should continue making budgetary provisions increasingly in strengthening the co-operative resources.[83]

Mirdha Committee (1964)

The Mirdha committee on co-operation was appointed by the government of india to suggest measures for the proper development of the Co-operative movement by eliminating non-genuine societies and vested interest. It submitted its report in August 1965. Some of the recommendations are given below.

(a) The principle of open membership must be enforced in all societies except that of traders.

(b) Elections to the managing committee should be held after 2 or 3 years. Election should be managed by the general body itself and not by the Registrar.

(c) Government should come to the assistance of Co-operatives in a big way. Apart from continuing the policy of state partnership, they should allow the local bodies and statutory corporation to use Co-operative banks as their bankers. The system of deposit insurance should be introduced in the Co-operatives also to ensure public confidence.

(d) Regular audit must be enforced by a penal provision if necessary, and audit machinery should be strengthened.[84]

Venkatappiah Committee (1969)

All-India Rural Credit Review Committee was set-up under the chairmanship of Sri.B.Venkatappiah to suggest measures for the reorganisation of rural credit. It suggested:

1. The establishment of an Agricultural Credit Board.
2. Setting up of a Small Farmers Development Agency in each of a number of selected districts throughout the country.
3. Creation of a Rural Electrification Corporation for the benefit of under-developed areas with an agricultural potential.
4. Formulation of a more active and much bigger role for the Agricultural Refinance Corporation.
5. Adoption of various measures for ensuring the Agricultural Refinance Corporation and
6. Adoption of various measures for ensuring the timely and adequate flow of credit for agriculture through co-operatives and commercial banks.[85]

Santhanam Committee (1969)

The committee on co-operation headed by Santhanam, in its report in 1969, said that, "The scale of cultivation finance should include a reasonable amount towards the consumption expenses of the members family. It further recommended that village societies should be empowered to pursue action under the land Revenue Recovery Act also to drive up recovery measures. In order to augment resources of the village societies, the committee recommended the margin between its lending rate to members and its borrowing rate from the central bank should be three percent.[86]

Farmers Service Societies (1971)

The National Commission on Agriculture, in its interim report in December 1971 recommended for the establishment of a new type of organisation at the base-level, called the Farmers Service Societies in order to

(i) To provide all types of credit, technical guidance and a full package of services especially to small farmers to develop their farms in integrated manner;

(ii) To cover effectively a large area of operation, say a block (or) population of 10,000 so that it could function as a viable unit; and

(iii) To provide for a two-third representation to enable the weaker sections to control society. Such societies could be either financed by commercial banks (or) by co-operative banks. It was accepted and a scheme was put into operation since 1973-93 in almost all the states by forming large-sized Adivasi Multi-purpose Co operative Societies.[87]

Bawa Committee (1971)

A special study group, under the chairmanship of Bawa (1971), recommended the organisation of large-sized Adivasi multi-purpose co operative societies in tribal areas as the bottom level structure.[88]

All-India Debt and Investment Survey: (1971-72)

This survey, undertaken by the RBI revealed that Rs.3,848 crores were the total debts of the rural households as on June 1971. This accounted for an increase of 96.73 per cent in the total rural debt over the position in 1961. Rs.3374 crores were the total debts of the cultivators in 1971. This represented an increase of 102.04 per cent as compared to their total debts in 1961. In the report, it was found that, "Although the non-institutional agencies still occupy a dominant position in the supply of credit to the rural households, their predominance is steadily declining while the institutional agencies are forgoing ahead. Thus, debt owed to non-institutional agencies formed 68 percent of the total debt of cultivators in 1971 as against 82 per cent in 1961.[89]

Hazari Committee (1975)

In 1975, RBI appointed a committee, under the chairmanship of R.K.Hazari for studying the feasibility of integrating the two wings of the co operative credit structure in the country. Noticing certain imbalances in the growth of the two wings of the credit structure, the committee felt that the co operative credit system can fully play its expected role if it is enabled to provide credit facilities in an integrated manner. Integration of the co-operative structure was also perceived to benefit the farmers by providing them with one contact point. The committee concluded that integration would prove to be beneficial from the angle of credit institutions.[90]

Kamath Working Group (1976)

The Reserve Bank of India constituted a working group under the chairmanship of C.E.Kamath in August 1976 to study the problems arising out of the adoption of multi-agency approach in agricultural financing and make appropriate recommendations. The working group submitted its report in April 1978 and its recommendations can be summed as:

"Considering the organisational potentialities of the co-operatives they are seen as the ideal institution to serve the credit needs of the agricultural sector".[91]

Sivaraman Committee (1979)

The committee was asked to review arrangements for institutional credit, for agriculture and rural development. It was headed by Sivaraman and the interim report was submitted in November 1979. The committee in its final report submitted in January 1981, had made several recommendations for strengthening the rural credit system in the country.

1. In the first place, the committee recommended, the grant of loans to the poorest of the rural population, i,e., the landless

labourers, rural artisans and small farmers with holding of less than half an acre for purposes like marriages, religions ceremonies, education and medical expenses. The committee estimated the demand of this group at Rs.170 crores annually such loans should be provided by the co-operative banks and the government.

2. The committee recommended grant of loans for similar purposes to marginal farmers holding half an acre to 5 acre of land. The committee estimated the demand of this group at Rs.125 crores, annually. Such loans should be provided by the co-operative and the scheduled banks.
3. In order to enable the poorer borrowers to return these loans, the states should work out schemes for their employment.[92]

Ojha Committee

The RBI appointed a high level standing committee under the headship of Ojha to review the flow of institutional credit to the rural sector. The terms of reference of the committee were to review and assess the working of co-operative credit agencies.[93]

Khusro Committee (1989)

RBI appointed, a committee headed by A.M.Khusro in 1989 for a review of the agricultural credit system in India. Khusro pointed out that the incidence of overdues in the agricultural credit system has been increasing over the years and has been steadily eroding the financial soundness of the system. It is very closely connected with the complex nature of agricultural economy and unless the basic issues involved in agricultural sector are tackled effectively, it will be difficult to find a lasting solution to the problem of overdues.[94]

Narasimham Committee (1991)

Shri M.Narasimham, was appointed Chairman of the committee on the Financial system in India by the Reserve Bank of India in 1991. He pointed out that the rural banks operations continued to dominate the rural areas and whose business would be predominantly engaged in financing of agriculture and allied activities. One of the major achievements of the banking system in the period following nationalisation has been the substantial branch expansion into the rural area and the increase in provision of credit to agriculture and allied activities.[95]

Jagdish Capoor Report (1999)

The Central Government constituted a Task Force in April 1999, under the chairmanship of Shri. Jagdish Capoor to study the functioning of rural Co operative credit institutions and to suggest a package of measures for their strengthening. A summary of the recommendations contained in the report submitted to the Central Government on July 24, 2000 is provided below.

(a) In view of the limited resources of co-operative banks, the Task Force has emphasised the need for strengthening the resource base, especially the capital.

(b) The Co-operative banks should work like professional organisations on sound managerial systems.

(c) The Task Force emphasized diversification of business products as the prime need at all levels in the co-operative credit institutions. The diversified avenues may include, interalia, housing loans, consumer loans, consortium financing, financing of services sector, distribution of insurance products, etc. The Task Force recommended that the co operative banks may be permitted to lend up to 10 per cent of their deposits outstanding as at the end of the previous year for commercial and technology intensive projects outside the co-operative fold.

(d) The Task Force is of the view that continuance of the existing three-tier structure in the short-term co-operative credit structure in bigger states is generally necessary. Further, the integration of ST and LT structures into a 'Single Window' organization may be on advantageous proposition.

(e) Mutual Assistance Fund (MAF), may be set up at the state level by contributions from co-operative institutions in the state for rendering assistance and providing soft loans to weak primary units to enable them to overcome temporary difficulties.

(f) The Task Force suggests that the provisions of the existing DRT may be made applicable to co-operative banks also where loan size is more than Rs.1 lakh so as to expedite recovery of chronic overdues. There is a need to evolve compromise/

settlement procedure for closing of long pending overdue loans.

(g) Audit of co-operative institutions should be conducted on a regular basis.

(h) The Apex co-operative banks and CCBs may be advised to disclose certain critical information in their balance sheets like movements in NPAs, provisions, return on assets, business per employee, profit per employee, etc.[96]

The entire discussion on the literature available in the field has shown that the co-operative movement has definitely made a marked impact on the agriculture development by providing institutional credit at a subsidised rate of interest. However, due to various constraints and in-built inabilities of the co-operative movement, the movement has not succeeded in eliminating non-institutional credit. Various economic and political factors along with frequent floods and droughts have made the farmers to abstain from repaying the debt. There is wilful default as well. Hence the very viable existence of co-operative credit movement is now at cross-road. Either there has to be a recovery of loans issued or they have to go dormant. The globilisation and financial reforms may put this subsidised agency at a disadvantageous position and the movement can succeed if only overdues are eliminated and these institutions stand on their own legs by mobilising savings.

REFERENCES

1. Bhalerao, M.M., Malati and Kavidra Singh, "Regional inequalities in long-term institutional credit -A study of land Development Bank", *Indian Journal of Agricultural Economics*, Vol.XLIII, No.3, July-September, 1988, p. 432.

2. Tyagi, R.B, *Recent Trends in the Co-operative Movement in India*, Asia Publishing House, New York, 1968, p. 64,71.

3. Mitra, A. and Sain, K., "Financing of Agriculture by Co-operative Banks in Burdwan district", *Indian Co-operative Review*, Vol. XX, No.1, July, 1982, pp. 25-31.

4. Bhagawan Rao and Kanakaiah, "Financing of Borrowers by Farmers Service Co-operative Societies - A case study", *Indian Co-operative Review*, Vol.XXI, No.1, July, 1983.

5. Muthur, B.S., *Co-operation in India,* Sahitya Bhavan, Jaipur, 1983, p. 60.
6. Hanumanthaiah, C.V., Raju, V.T. and Naidu, M.R., "Temporal Variation in Economic and financing Aspects of cotton farms-A case study", *Indian Journal Agricultural Economics,* Vol.XLIII, No.3, July-September, 1988, pp.427-428.
7. Baidyanath Misra, *Co-operative movement in India,* A.P.H. Publishing Corporation, New Delhi, 1997, p.159.
8. Krishna Kumari, A.B., *Co-operative Banking for Agricultural Development,* Deep and Deep Publications, New Delhi, 1998, p.23.
9. Mudgal, M.K., "Co-operatives at Cross-Roads", *Kurukshetra,* Vol.46, No.8, May, 1998, p.13.
10. Atchi Reddy, M, and Prabhakar, "Rural Credit in Karim Nagar District", *Indian Journal of Agricultural Economics,* Vol.XLIII, No.3, July-September, 1988, p.426.
11. Mamoria. C.B. and Saksena, R.D. *Co-operation in India,* Kitab Mahal, Allahabad, 1977, p.246.
12. Rajkishore Panda, "Credit Finance of Primary Agricultural Co-operative Societies Among Farm Households in Orissa, *Indian Journal of Agricultural Economics,* Vol.XLIII, No.3, July-September, 1988, p.437.
13. Ratnaparkhe, A.N. and Pagire.B.U., "A study into the Appropriateness in the scale of Finance by the Co-operatives in Maharashtra State," *Indian Journal of Agricultural Economics,* Vol.XLIII, No.3, July-September, 1988, P.490.
14. Beohar and Khare, "A study of the Recovery and overdues of Agricultural loan issued by Bishnoo Datta Central Co-operative Bank, Branch Umaripan, Jabalpur district (M.P)", *Indian Journal of Agricultural Economics,* Vol.XLIII, No.3, July-September, 1988, pp. 426-427
15. Suryawanshi, S.P., "Credit Requirement and its Gap on the Farm in Bhima Command Area of Western Maharashtra," *Indian Journal of Agricultural Economics,* Vol.XLIII, No. 3, July-September, 1988, pp. 424-425.
16. Dinesh, K. Marothia, "Production Credit Management in Rice Farming System of Chhattisgarh Region", *Indian Journal of Agricultural Economics,* Vol.XLIII, No.3, July-September, 1988, pp.434-435.
17. Srinivasan, S., "Long-Term Credit in Anna District Tamil Nadu-A case study", *Indian Journal of Agricultural Economics,* Vol.XLIII, No.3, July-September, 1988, p.419.
18. Gangadharan, T.P., "Infirmities in Rural Credit Policy-A case study of Farmers Loans in Kozhikode District", *Indian Journal of Agricultural Economics,* Vol.XLIII, No.3, July-September, 1988, p.402.

19. Amrik Singh Saini, Raj Vir Singh and Patel, K.K., "Rural Credit Management through integrated crop and milk production", *Indian Journal of Agricultural Economics,* Vol.XLIII, No.3, July-September, 1988, p. 427.
20. Munidoraswamy Naidu, K and Surendrababu, K.B., "Rural Credit Policy: Need for Transcending Credit Needs by creating debt Capacity", *Indian Journal of Agricultural Economics,* Vol.XLIII, No.3, July-September, 1988, pp.417-418.
21. Vaidya. C.S, Sikka.B.K. and Swarup, R, "Cost of Credit in Hill Agriculture: A Spatial Analysis," *Indian Journal of Agricultural Economics,* Vol.XLIII, No.3, July-September, 1988, pp. 439-440.
22. Alamelumangai, R, "Can Co-operatives Survive and succeed in the context of Economic liberalisation"? *The Tamil Nadu Journal of Co-operation,* Vol.87, No.5, August, 1995, p.37.
23. Satish Verma and Vijay Chopra, "Demand for Bank Credit in the Agricultural sector in India", *Indian Journal of Agricultural Economics,* Vol.XLIII, No.3, July-September, 1988, p.438.
24. Pandey, R.K. and Ashok Kumar, "Study of Growth and Disparity in Flow of Co-operative Credit in India," *Indian Journal of Agricultural Economics,* Vol.XLIII, No.3, July-September, 1988, p.416.
25. Pandey, R.K. and Guglani, P, L., "Trends in Recovery overdues and outstanding position of Agricultural Loans, Ranch District (Bihar)", *Indian Journal of Agricultural Economics,* Vol.XLIII, No.3, July-September, 1985, p. 434.
26. Ramachandra Bhat and Aruna Rao, "The Evaluation of the Growth of Primary Agricultural Credit Co-operative Societies in Karnataka 1965-1978", *Indian Co-operative Review,* Vol.XIX, No.3, April, 1982.
27. Sarker, S.C.., Mitra, A and Mukhergee, A.K., "Institutional Credit in Developed and Underdeveloped Villages of Naidu District West Bengal an appraisal," *Indian Co-operative Review,* Vol.XXIV, No.2, October, 1986.
28. Sharma, M.L, and Teweri, "Effect of Alternative Production Credit Allocation Policies on growth and Equity in a Community Development Block", *Indian Journal of Agricultural Economics,* Vol.XLIII, No.3, July-September, 1988, pp.416-417.
29. Gayal, S.K. and Pandey, "Trend Analysis of Primary Agricultural Co-operative Credit and Service Society in Haryana," *Indian Co-operative Review,* Vol.XXVI, No.1, July, 1988.
30. Ganesh, P., Gupta and Sudha Ran Sahu, "Strategy for promotion of loyalty in the Co-operative System," *Co-operative perspective, Journal of Co-operative Management,* Vol.34, No.4, Jan-March, 2000, pp.5-6.

31. Op.cit. Mamoria, C.B. and Saksena, R.D., p.221.

32. Mathur, B.S., *Co-operation in India,* Sahitya Bhavan, Agra, 1980, pp. 59-64.

33. Proceedings of the Fifteenth and Sixteenth Meetings of the Agricultural Credit Board, Reserve Bank of India, Bombay, 1981, p. 222.

34. Sankaraiah, G. and Munidoraiswamy Naidu, K., "Financing Small farmers by Co-operatives", *Indian Co-operative Review,* Vol.XXI, No.1, July, 1983.

35. Rayudu, C.S., "Participation of Co-operatives in Small farmers Development - A case study", *Indian Co-operative Review,* Vol.XXI, No.4, April, 1984.

36. Satyanarayana, I, Ch. Rushyendra Rao and Venkatesa Rao, "Subsidy to Identified Beneficiaries by FSS-A case Study," *Indian Co-operative Review,* Vol.XXIII, No.3, January, 1986.

37. Viswanadha Rao and Mohandoss, "Co-operative credit certain Dimension and problems", *Indian Co-operative Review,* Vol.XXIII, No.3, January, 1986, p. 257.

38. Singhal, C.D., "Evaluation of Role of Primary Credit and Service Societies (PACS) in Haryana-A case study," *Indian Co-operative Review,* Vol.XXVI, No.1, July, 1988, pp.7-12.

39. Dandekar, M.M., *The Indian Economy 1947-92 Agriculture,* Vol.I, Sage Publications India Pvt. Ltd., New Delhi, 1994, p. 367.

40. Mamoria, C.B., *Agricultural problems of India,* Kitab Mahal Pvt. Ltd., Allahabad, 1995, p.421.

41. Gopal Lal Jain, *Rural Development,* Mangal Deep Publications, Jaipur, 1997, p. 236, 237.

42. Chandra Prasad, J. and Visweswara Rao, K.V.V.S., "Co-operative Credit Agencies": An Evaluation, *Kurukshetra,* Vol.46, No.11, August, 1998, p.34.

43. Verma, S.K, "Rural Development through Co-operatives", *Kurukshetra,* Vol.47, No.8, May, 1999, pp. 31, 32, 38.

44. Balishter and Chauhan T.R., "A study of Co-operative Finance in Agriculture". *Indian Co-operative Review,* Vol.XIX, No.3, April, 1982, p.52.

45. Chauhan, T.R., Balishter and Rajkumar, "Availability and Extent of Utilisation of Credit in Agriculture," *Indian Co-operative Review,* Vol.XX, No.1, July, 1982, pp.47-48.

46. Radhakrishnan, V. and Mukundan, "Supply and Utilisation of Short-term Co-operative Agricultural Credit in Palghat district", *Indian*

Journal of Agricultural Economics, Vol. XLIII, No.3, July-September, 1988, p.429.

47. Gupta and Verma, "Role of Co-operative Credit in Increasing Agricultural Production in M.P", *Indian Co-operative Review,* Vol. XXII, No. 1, 1986.

48. Rajput, S.S., "Rural Credit and its cost - A Case study", *Indian Co-operative Review,* Vol.XXIII, No.3, January, 1986, p.301.

49. Vaikunthe, L.D., "Utilisation of Credit in Agriculture - A case study," *Indian Co-operative Review,* Vol.XXV, No.4, April, 1988, pp. 361-368.

50. Kaliya Perumal, P. and Loganathan, P., "Agricultural Credit in Elakurichi Village a Micro Analysis", *Indian Co-operative Review,* Vol. XVII, No. 2, October, 1989.

51. Baidyanath Misra, *Co-operative Movement in India,* A.P.H. Publishing Corporation, New Delhi, 1997, p. 181.

52. Kalyan Bhattacharya, "Financial intermediation and Rural Development : Role of Co-operative Credit in Agricultural Development", *The Indian Economic Journal,* Special Conference number for 1978, Part-II, Vol. 26, No. 2, October-December, 1978, p. 13.

53. Adinarayana, S. and Lakshminarayan, M, "Impact of Institutional Production Credit on farm Productivity in Kasimkota panchayat Samithi of Visakhapatnam District (A.P)", *Indian Journal of Agricultural Economics,* Vol.XLIII, No.3, July-September, 1988, pp. 430-431.

54. Sami Uddin and Mahfoozur Rahman, *Co-operative Sector in India-After Independence,* S.Chand and Company Ltd., New Delhi, 1983 p. 40.

55. Srivastava, G.C. and Lal, K.K., "Economic Rationale of Credit use on small Farms - A case study in Bihar", *Indian Journal of Agricultural Economics,* Vol.XLIII, No.3, July-September, 1988, pp. 431-432.

56. Singh, S.K., Ram Iqbal Singh and Singh, G.N., "Impact of Rural Credit through Co-operative in Agricultural Development in Eastern Uttarpradesh," *Indian Journal of Agricultural Economics,* Vol. XLIII, No.3, July-September, 1980, p. 422.

57. Lal, R.C., and Lavania R.P., "Impact of Co-operative Credit on Agricultural Production and Income", *Indian Co-operative Review,* Vol. XXIII, No. 3, January, 1986.

58. Hemchand Jain, "Impact of Institutional Credit on Tribal and Non-tribal Farms in Mandla District of Madhya Pradesh", *Indian Journal of Agricultural Economics,* Vol.XLIII, No.3, July-September, 1988, p. 418.

59. Singha, R.P. and Sharma, B.M., "Role of Credit and Technology in Increasing farm Income Labour Employment - A case Study", *Economic Affair,* Vol.33, quarterly, March, 1988, pp. 43-44.

60. Bhatia, B.S., Verma, H.L., and Garg G. Mahesh, "Encyclopaedia of Co-operative Management", Vol. 3, *Co-operative Banking: Leavers of Rural Economy,* Deep and Deep Publication, New Delhi, 1994, p. 129.

61. Gulab Singh Azad, "Role of Co-operatives in Rural Development," *The Tamil Nadu Journal of Co-operation,* Vol.88, No.11, February, 1997, Chennai, p. 29.

62. Duhan, T.C. and Singh, "Institutional Credit for Dairying in Haryana State", *Indian Journal of Agricultural Economics,* Vol. XLIII, No. 3, July-September, 1988, pp. 425-426.

63. Bhosal, S.K. and Bangat, S.B, "A Study into the overdues of Co-operative Loans in Maharashtra", *Indian Journal of Agricultural Economics,* Vol.XLIII, No.3, July-September, 1988, p. 420.

64. Gangwar, A.L. and Kusam Aggarwal, "Borrowing and Repayment Performance of Farmers Pertaining to Institutional Loans in Kurukshetra District, Haryana", *Indian Journal of Agricultural Economics,* Vol. XLIII, No. 3, July-September, 1988, p. 443.

65. Vasant Desai, *A Study of Rural Economics-A Systems Approach,* Himalaya Publishing House, Bombay, 1983, pp. 516, 517.

66. Datta, T.N., "Analysis of Default in Organised Sector Credit - A study of minor Irrigation scheme in West Bengal," *Indian Journal of Agricultural Economics,* Vol.XLIII, No.3, July-September, 1988, p.428.

67. Julka, A.C., Soni, R.N. and Arora, S., "Profile and cost of Rural Credit: A case study", *Indian Journal of Agricultural Economics,* Vol.XLIII, No.3, July-September, 1988, p.440.

68. Raghu Ram, P. and Bhavani Devi, I, "An analysis of Tangible means to Reduce Incidence of Default in Commercial Bank Lendings," *Indian Journal of Agricultural Economics,* Vol.XLIII, No.3, July-September, 1988, pp.422-423.

69. Khusro, A.M., (Chairman), "A Review of the Agricultural Credit system in India", *Reserve Bank of India* (RBI), Bombay, 1989, p.535, 566, 567.

70. Ranga Reddy, *Agricultural Development Rural Credit and Problems of its Recovery,* Mittal Publications, New Delhi, 1990, p.73, 172.

71. Dutta.S.K., *Co-operative Societies and Rural Development,* Mittal Publications, New Delhi, 1991, p. 167.

72. Radhakrishnan, N. "Institutional Credit for Agricultural Sector - A case study of Mambakkam Primary Agricultural Co-operative Credit

Society: 1985-86 to 1992-93, M.Phil., Thesis, Department of Economics University of Madras, 1994, p.134, 138.

73. Anandharajakumar, P. and Arockiaraj, K. "Primary Agricultural Credit Societies Instrument for Rural Development", *The Tamil Nadu Journal of Co-operation,* Vol.87, No.8, November, 1995, p.8.

74. Hand Book of Indian Agriculture, Vikas singhal, vikas Publishing House Pvt, Ltd., New Delhi, 1995, p.17.

75. Rangarajan, C. "Issues in Rural Credit", *The Tamil Nadu Journal of Co-operation,* Vol.89, No.5, August, 1997, p.24.

76. Tamil Mani, B. and Manivel, S., "Repaying the Dues to the Co-operative Banks is the best culture for the growth of Co-operatives," *The Tamil Nadu Journal of Co-operation,* Vol.89, No.7, October, 1997, p. 24.

77. Krishna Kumari, A.B. *Co-operative Banking for Agricultural Development,* Deep and Deep Publications, New Delhi, 1998, p.25.

78. Report on trend and progress of Banking in India 1998-99, RBI, Supplement to *RBI Bulletin,* December, 1999, Mumbai, p.130.

79. Tyagi, R.B., *Recent Trends in the Co-operative Movement in India,* Asia Publishing House, Bombay, p. 14,15.

80. *Ibid.* p.119, 133.

81. Uduman Mohideen.K.S.S, *Institutional Credit and Agricultural Development,* Mittal Publications, New Delhi, 1991, p. 28.

82. Op. cit. Mamoria, C.B. p. 440.

83. Op.cit. Uduman Mohideen K.S.S., p.29.

84. Bedi, R.D., *Theory, History and Practice of Co-operation,* International Publishing House, Meerut, 1987, p. 201,202.

85. Op.cit. Mamoria C.B. and Saksena, R.D., pp. 214, 215.

86. Op.cit. Uduman Mohideen K.S.S., p. 35.

87. Op.cit. Uduman Mohideen K.S.S., p. 38.

88. Op.cit. Uduman Mohideen K.S.S., p. 41.

89. Op.cit. Uduman Mohideen K.S.S. p. 31.

90. Report on Trend and Progress of Banking in India (1998-99), Supplement to *RBI Bulletin,* December, 1999, Mumbai, p. 129.

91. Tyagi, B.P. *Agricultural Economics and Rural Development,* Jai Prakash Nath & Co., Meerut, 1990, p. 694.

92. *Ibid.* p. 687.

93. Op.cit. Uduman Mohideen K.S.S., p. 86.

94. Op.cit. Khusro, A.M. p. 535, 566, 567.
95. Narasimham, M (Chairman) "Report of the Committee on the Financial System in India", *RBI*, Bombay, 1991, pp. 69-75.
96. Report on Trend and Progress of Banking in India, 1999-2000, *RBI* Mumbai, pp. 121.122.

Chapter - 3

POLICY INTERVENTION: CO-OPERATIVE MOVEMENT IN WORLD AND INDIA

CO-OPERATIVE MOVEMENT: WORLD WIDE

The Co-operative movement today is compelled to develop more and more at the international level. In its spirit it knows no national boundaries, but the process of its development is varied depending upon natural, social, economic and political situations in the respective countries. Accordingly, it is necessary to take into consideration the circumstances under which the co-operatives are placed in South-East Asia. In most Asian countries, agricultural co-operatives were developed initially to save the farmer from the onerous terms of the moneylender.

As most countries in the region are predominantly agricultural, co-operative organisations are helping farmers to procure finances, market their supplies and obtain consumer goods for household requirements. In several countries, as a consequence of land reform measures,there has been an emergence of small holdings. Farm production can be augmented by the organisation of co-operative institutions by providing the required services.

The provision of credit to agriculture by co-operation in Europe and Asia is one of the oldest forms. For more than a century the names Raiffeisen and Schulze-Delitzch have been accepted as international technical terms originating from Europe. In Asia, for several centuries, various forms of saving and lending at the village community level are known to have existed in the region according to rural economic historians. Owing to the predominantly agricultural character of the economies of most Asian countries, and because of the initial emphasis on that sector, most co-operative movements show main development in the field of agriculture and especially in the provisions of credit. India has the largest number of primary co-operatives in existence among the Asian countries; of a total of around 332000 societies in 1960/61 a little over 70 percent were credit societies (around 95 percent of these being agriculture credit societies)[1].

Table 3.1
Formation of the First Co-operatives in Asian and African Countries

Asian Countries		Year of Formation	African Countries	Year of Formation
1.	India	1900	Egypt	1908
2.	Srilanka	1904	Kenya	1908
3.	Burma	1905	Botswana	1910
4.	Philippines	1910	Mauritius	1913
5.	Thailand	1917	Uganda	1913
6.	Indonesia	1920	Libya	1915
7.	Malaysia	1920	Senegal	1916
8.	Jordan	1922	Burundi	1921
9.	Syria	1934	Morocco	1922
10.	Nepal	1950	Ghana	1928
11.	S.Yemen	1957	Togo	1931
12.	Saudi Arabia	1961	Guinea	1940

Source: MASLENNIKOV. V. The Co-operative movement in Asia and Africa, Progress Publishers. Moscow, USSR-1983, p.15.

The Co-operative movement which saw the light of the day in Europe has spread throughout the world. "Co-operation" Dr.Fauguent writes, "is perhaps more widely distributed geographically than any other modern form of organisation".

France has played an important role in the promotion of agriculture co-operatives. The co-operative movement in sweden is divided into three distinct parts, viz., the agricultural co-operative movement, the consumers co-operative movement, and housing co-operative movement. Sweden is a small agricultural country. There is no organisational and structural connection between these three branches of the movement as each works independently.[2]

The genesis and the rise of a co-operative movement are greatly influenced by the historical conditions and the economic and social systems in which it operates. Co-operation in the U.S.A. developed faster and more extensively among farmers than among urban population. Thus the movement is predominantly in the agricultural sector. Germany and England exerted their influence on Japan in choice of co-operatives to be founded. The Japanese farming villages,realizing the important place of co-operation in farming, formed co-operative societies.

Germany is the birth place of the co-operative credit movement in the world. In India, the inspiration for the co-operative movement came largely from Germany. Mr.Fredick Nicholson, who was sent to Europe to study the working of agricultural banks and suggest measures for the introduction of co-operative movement in India, based his recommendations mostly on the pattern of credit societies in Germany. His important advice to Indian co-operators is Find 'Raiffeisen' and he gave importance to the Raiffeisen system of credit as a means of solving problems of indebtedness in India.[3]

CO-OPERATIVE MOVEMENT IN INDIA

Co-operative organisations often arise on the initiative of the working people themselves.[4] "Co-operation" is derived from the latin word `co-operari'. `Co' means `with' and `operari' means `to work'.[5]

Co-operation as an organisational feature of the community and the farm life in india is as old as the village community. Interdependence or co-operation has always been the key note of the Hindu family; joint in food, worship, and estate. During very early days of the Maurya Dynasty, Chanakya in his Arthshastra remarked, "Who ever stays away from any kind of co-operative understanding shall send his servants and bullocks to carry on the work, he shall have a share in the expenditure but not in the profit". Vedas and the Law of Manu also mention craft guilds with banking as a feature of their activity. The village has been a sort of co-operative body since ancient days. In feudal days, although the villages were controlled by feudal landlords, they were co-operatively incharge of their own management.[6] Co-operation is the backbone of economic activities and social progress. Co-operation leads to social and economic harmony. Co-operation is that form of economic organisation in which persons wilfully and voluntarily pool their resources on a basis of equality for advancement of their economic interests.[7] Co-operation is a special mode of doing business and at present is applied for the solution of many economic problems. Co-operation is harnessed to almost all forms of economic activity, though co-operation was introduced in this country as a remedy for rural indebtedness. The co-operative movement is a symbol of hope to the down-trodden and embodies self-help and mutual aid among its votaries producing a sense of self-respect and brotherhood amongst them.[8]

In most of the countries of the world an attempt has been made to develop institutional credit for agriculture on co-operative lines. The co-operative form of organisations are considered best for providing credit to the farmers. The increase of agricultural production is ultimately bound up with organisation of agricultural credit, which places at the disposal of agriculturist the necessary capital on favourable terms. Towards the end of the last century, the peasantry was in the grip of the Sahukar and the rural debt was daily mounting. The condition of the agriculturists was really deplorable. The joint family system was breaking down. As a result of increase in population and the absence of rapid industrialisation more and more people depended upon agriculture resulting in too much pressure on land. The land got increasingly divided and sub-divided. Farmers who were once prosperous became petty peasants with tiny holdings of a few acres and were finally reduced to the status of landless labourers. They were in the clutches of the moneylenders.[9]

A cultivator is in almost perpetual need of credit both for production and consumption. Traditionally, his needs were met by the landlord, traders, money-lender nexus involving usurious interest rates and other exploitative practices. Official cognizance of his needs goes back to 1793, when the system of taccavi loans was introduced providing loans at low rates of interest for agricultural improvements, mainly for the digging of wells. During the rule of Lord Mayo (1869-1872), the land Improvement Act 1871, was passed. It was amended in 1876 and repealed in 1883 when a more comprehensive piece of legislation was enacted, the Land Improvement Loans Act 1883, 'to consolidate and amend the law relating to loans by the government for agricultural improvements. Taccavi Loans were to be repayable in instalments which could be extended over a period of 35 years depending upon the durability of the work (or) the purpose for which the loan was advanced. Further, the Agriculturists Loans Act was passed in August 1884 to provide the advance of loans to the owners and occupiers of arable land for the relief of distress, the purchase of seed, cattle (or) any other purpose not specified in the Land Improvement Loans Act 1883, but connected with agricultural objects. Government finance continued to be channelled under the provisions of the Acts of 1883 and 1884. Some special regulations for advancing loans for specific purposes like natural calamities, pump-sets and agricultural machinery were also passed by a number of states/provinces.[10]

The term village banks was first used by Nicholson, in his report of 1895. The Deccan riots created the necessary background for thinking about banks for the famine struck, indebted farmers. The condition of the farmers, who participated in the Deccan riots was too bad to lead a normal life, because of their indebtedness. The landlords and Savakars were the major factor responsible for the farmer's indebtedness. The idea of establishing agricultural banks has been a much discussed subject ever since the Deccan riots. The drought situation and the extravagant expenses in the social and cultural activities of the agriculturists resulted into the borrowing from the moneylenders and traders. Their exploitation was the outcome of the riots. Sir Wedderburn and Justice Ranade were the first proponents of agricultural banks. In 1892, Government of Madras directed Sir Frederick Nicholson to study agricultural banks in the province. Mr.H.Dupernex supported the concept of village banks as peoples bank. He discovered the historical fact that, even before Schulze Delitsch, Raiffeisen, Luzzatti and Well Emberg, the advantages of association on co-operative lines had been perceived by the great philosopher and warrior king Frederick the Great of Prussia. F.Nicholson suggested in his very valuable report on "Land and Agricultural Bank", the kind of credit, primarily needed by the ryots of this presidency i.e., continuous small loans for current needs, including maintenance, cultivation expenses, purchase of stock, payment of revenue, rent and the like, and secondly, long-term loans for permanent improvements. Whatever may be the necessary additional capital for improvements be, a more pressing and immediate need is for cheap and safe credit in current transactions.[11]

Before the introduction of the official co-operative movement in India in 1904 steps were taken by the government to alleviate the sufferings of the farmers by the provision of Taccavi loans. Taccavi loans for land development/improvement (for long-term purposes) were given under the Act of Land Improvement Loans Act of 1883 and loans for cultivation purposes (short-term) were given under Agriculturists Loans Act of 1884.

Under the Agriculturists Loans Act, State Governments were empowered to make rules as to the loans to be given to farmers. Sir Stuart Baylay emphasised that the Land Improvement Act 1871 had been unsuccessful because it still made it necessary for the cultivator to resort to money-lenders. The Agricultural banks contemplated at the time of passing the Land Improvement Loans Act never came into existence and Agriculturists Loans Act had practically no effect other than that of extending the system of Taccavi. One of the reasons for the failure of Taccavi loans was that they were handled by the Revenue Department which was concerned with revenue collection. The famine commission in 1880 had recommended the administration of the land improvement loans through the agricultural department. The main reason for the introduction of the co-operative movement in India was the failure of Taccavi loans.[12]

The Land Improvement Act of 1883 and the Agriculturists Loans Act of 1884 as well as other relief measures were found to be highly incommensurate in coping with the stupendous and complex problem of rural indebtedness. The experience of the administration was that the relief measures tried upto that time were disappointing; they had done little substantial good as the Famine commission of 1901 has remarked. The officials of the Government of India, who had studied working of agriculture banks of Egypt, were strongly of the view that India must go in immediately for co-operative experiments on similar lines.[13]

Co-operative movement was launched in India, with the passing of Co-operative Societies Act in the year 1904. Under this Act, Co-operative thrift and credit societies were registered. Since the economy of India has been predominantly an agricultural economy, co-operative societies were established to ameliorate the economic condition of farmers by way of providing cheap and adequate finance to them and saving them from the clutches of moneylenders.[14] The recommendations of Frederick Nicholson,

Dupernox and Sir Edward Law provided the basis for the bill which was passed in 1904, as the Co-operative Credit Societies Act.

The co-operative movement in India came from the government. India is not unique in having had the original impetus to co-operation come from the government officials, or in having the movement receive government encouragement and even support, but there has been in India a degree of dependence upon official initiative, direction and control that is probably without parallel. A co-operative society was started in Punjab as early as 1891 for controlling the common land of the village for the benefit of the co-shares, and functioned until 1922, when the land was partitioned. Agricultural banks were introduced in Mysore in 1894, each 'to be an association of landholders on strictly co-operative principles.

In Bombay presidency it was ruled in 1921 that, in villages where co-operative societies were situated, advances under the Land Improvement Loans Act might be made only through them. These provisions for loans to agriculturists not having met the situation, other plans were considered. A committee meeting in Calcutta in December 1900 considered the introduction of co-operative credit societies, reaching the conclusion that societies on Raiffeisen lines might prove suitable. The Madras presidency, even before the movement was officially launched, had organised co-operative undertakings in the form of the indigenous nidhis (or) mutual loan associations not unlike the provident fund and friendly societies of Europe.[15]

The idea of co-operation received support from the Famine Commission of 1901. It has been strongly recommended the introduction of credit association. The result was the passing of the Indian Co-operative Credit Societies Acts, 1904.

The co-operative movement in India had passed through the following stages.

The First Stage of the Movement (1904-1911)

The Co-operative Credit Societies Act was passed on 25th March 1904. The introduction of the Co-operative Credit Societies Act in 1904 marked the beginning of the co-operative movement in India. This laudable measure was hailed as, "a turning point in economic and social history" by Henry Wolff, the great co-operator and "the way from poverty to plenty" by Sir Daniel Hamilton. The object of this Act, as stated in the preamble, was to encourage thrift, self-help and co-operation amongst agriculturists, artisans and persons of limited means.[16]

Second Stage (1912-1918) Period of Hurried Expansion

The defects of the 1904 Act were remedied in 1912 when another Co-operative Societies Act was enacted. With the passing of this Act, the movement entered on the second stage of its progress.

The essential features of this Act are the following :

(a) Under this Act any society, credit (or) otherwise may be registered which has as its object the promotion of the economic interests of its members in accordance with the co-operative principles.

(b) A federal society like the central bank could be registered.

(c) Unless otherwise directed by state governments,

   (i) the liability of central societies shall be limited, and

   (ii) the liability of rural credit societies shall be unlimited.[17]

Third Stage (1919-1929)

Under the reform of 1919, co-operation became a transferred subject, under the charge of a minister, in each state. The total number of societies of all kinds in India in 1923 was nearly 56,136. Of these the agricultural co-operative credit societies alone numbered nearly 49,118. Thus it is clear that the movement has developed almost entirely in the direction of agricultural credit societies. At the time of the appointment of the Royal Commission on Agriculture in 1926, the number of agricultural credit societies in India was about 65,101. The Royal Commission on Agriculture, with Mr.Calvert as Chairman, was set up during this period. The commission sounded a note of warning in its famous words, "If co-operation fails, there would fail the best hope of rural India". The Townsend Committee was appointed in Madras as early as 1927, the King Committee of the Central Provinces and Berar in 1922, the Oakden Committee of U.P. in 1925, the Calvert Committee of Burma in 1928.[18]

The Act of 1919 gave great stimulus to the movement. Its success was measured more by its quantity than by its quality. The number and membership of agricultural credit societies increased three times while their working capital increased four times between 1918-19 to 1928-29.

The rapid growth of movement between 1919 and 1930 is characterised by Mr.Ramdas Pantulu as the period of "Unplanned expansion".

Fourth Stage: (1929-1939)

In 1929 the world economic depression set in and gave the movement a rude shock. The main problem that faced by various committees was to find out a general method of re-examination of loans due to primary societies by individual members, with a view to scaling them down for repayment in easy instalments over a period of 20 years. The large acreage of land that had earlier come into the possession of primary societies and banks in lieu of debts had to be returned on a sort of hire-purchase system. Some of the societies were advised to write off bad debts out of their owned funds, and if necessary, by resorting to the floatation of debentures. From 1935 to 1939, there were three main developments in the field of co-operation. Like,

(a) In 1935, the Reserve Bank of India, was established and its Agricultural Credit Department was charged with the duty of studying various problems relating to agricultural credit.

(b) In 1936, an All India Conference of Registrars of Co-operative Societies stressed the need for development of long-term credit through co-operative land mortgage banks.

(c) Finally, the concept of primary co-operatives came to be increasingly accepted.[19]

In 1919, the responsibility for co-operative legislation was shifted to states/provinces which gave adequate stimulus to provincial states to enact independent co-operative legislation.

(a) Madras Co-operative Societies Act, 1934.

(b) The Bombay Co-operative Societies Act, 1935.

(c) Bihar & Orissa Co-operative Societies Act, 1935.

(d) Bengal Co-operative Societies Act, 1940.

(e) Assam Co-operative Societies Act, 1949.

These act paved the way for the formation and development of a wide variety of co-operative institutions across the country.

Fifth Stage (1939-1947)

During this period the co-operative movement made rapid progress as a result of the Second World War. The farmers gained heavily owing to an increase in the prices of agricultural commodities. A number of consumer co-operative stores were established. During this period the number of societies increased from 122000 in 1938-39 to 172000 in 1945-46. The membership and working capital during the same period rose from 53.7 lakhs to 91.6 lakhs and Rs.106.47 crores to 164 crores respectively. The Government of India appointed in 1944, a committee known as the policy committee on agriculture, forestry and fisheries to report about the indebtedness of agriculturists and suggest methods to reduce the same. The committee, with Prof.D.R.Gadgil as chairman, reported that co-operative developments were not uniform in the different provinces and hence they suggested the establishment of an agricultural credit corporation for each province except those where co-operation was thriving. The co-operative planning committee, with Sri.R.G.Saraiya as chairman, submitted its comprehensive recommendations on all aspects of co-operation in 1946. The 15th conference of Registrars of Co-operative Societies, which met in 1947, approved many of the recommendations of Co-operative planning committee. There were about 84,000 primary societies with a membership of 31.75 lakhs, paid up capital of Rs.14.43 crores, deposits of Rs.3.11 crores and a working loans to tune of about Rs.9.03 crores in 1947.[20]

The period 1904-1947, was nurtured by the British Government, People's initiative was hardly forthcoming. The movement depended mainly on official guidance and support.

Growth of Co-operative Sector: After Independence

The First Five-Year Plan (1951-56) clearly expressed preference for the co-operative organisation of the economic activities of the people, especially in regard to agriculture, marketing, cottage and processing industries. The First-Plan emphasized that provision of sufficient and timely credit at fair rates of interest was to be considered as an integral part of the plan. The plan pointed out the need to evolve a comprehensive and integrated policy of agricultural credit. The history of agricultural development in all advanced countries show that such an integrated system of credit laid the foundation for agricultural development and prosperity.[21]

Under the Second Five-Year Plan (1956-61) : The role of co-operatives in the formation of capital and in the improvement of rural life in general has been emphasised by the planning commission. Reliance has been placed on the co-operative movement as one of the sources of financing agriculture. It was proposed that during the second five-year plan the active membership of primary agricultural credit societies should be raised from 5 to 15 million and the amount advanced through the co-operative movement in short-term loans should be raised from Rs. 30 crores to Rs.150 cròres, in medium-term loans from Rs.10 crores to Rs.50 crores and in long-term loans from Rs.3 crores to Rs. 25 crores.[22]

Third Five-Year Plan (1961-1966)

The third plan recognized co-operation as a vital factor for social stability and economic growth and visualised wider dimensions for it in the country. The third-plan pledged to the values of socialism and democracy, co-operation should become progressively the principal basis of organisation in many branches of economic life, notably in agriculture and minor irrigation, small industry and processing, marketing distribution, civil supplies, rural electrification, housing and construction, and the provision of essential amenities for local communities. Before set-up the fourth Five-year plan, Government of India had declare three annual plans (1966-67, 1967-68 and 1968-69). The major targets, for 1967-68 was to achieve a total membership of 40 million in PACBs by the end of June 1968. During the period 1966-68 an important development in India was the implementation of the New Agricultural Strategy, which aims at self-sufficiency in food by 1970-71. The All-India Rural Credit Review Committee, which was constituted by the Reserve Bank of India in 1966, was also active. It has recommended, among other things, the establishment of a Statutory Agricultural Credit Board in the RBI, the setting up of a Small Farmers Development Agency (SFDA) in selected districts, establishment of Rural Electrification Corporation and the expansion of the role of the Agricultural Refinance Corporation.[23]

The necessity to consolidate the movement and adopt measures to revitalise the same was felt at the conference of the state Chief Ministers and Ministers of co-operation held at Madras in 1968.

During the Fourth Five-year Plan (1969-74) it was stated that "It is important for planned development to bring about growth of co-operatives in all parts of the country to ensure the co-ordinated operation of various types of co-operative organisations". During the Fourth-Plan, it was decided:

(i) to reorganise the primary credit societies by making them viable units;

(ii) to rehabilitate and reorganise weak district central co-operative banks;

(iii) to finance primary societies directly by the concerned Apex Co-operative Banks;

(iv) to reduce overdues;

(v) to strengthen the agricultural credit stabilisation funds;

(vi) to provide trained and competent staff for the supervision of primary societies;

(vii) to encourage deposits, co-operative banks were required to open more branches in the rural areas.

The major objective of the Fifth Five-Year Plan (1974-1979) was to build up a strong and viable co-operative sector with special emphasis on the needs of cultivators, workers and consumers.[24]

In March 1979, the RBI appointed a committee to suggest improvements in the existing arrangements for institutional credit for agriculture and rural development (CRAFICARD). The committee noted that problems of agricultural credit had not only grown in complexity and size but had also merged with the larger tasks of rural development, and recommended the setting up of a new apex bank - the National Bank for Agriculture and Rural Development (NABARD). NABARD was to take over from the RBI the over issuing of the entire rural credit system including credit for rural artisans and village industries.

The Sixth Five-Year Plan (1980-85) has been assigned an important role in the rural development, especially agricultural sector. A clearly conceived action programme has been drawn up for strengthening of primary societies, so that they are able to effectively act as multi-purpose units catering to diverse needs of their members. There are some special programmes of rural development like IRDP, DPAP national grid of rural godowns in which co-operatives have been assigned an active role, other programmes like co-operative credit, marketing, processing and storage, consumer co-operatives, co-operative training as education etc, are directly concerned with the co-operative sector. The Sixth Five-Year Plan (1980-85) endorsed the setting up of NABARD which was established by an Act of Parliament in July 1982, for providing credit for the promotion of agriculture, small-scale industries, cottage and village industries, handicrafts and other rural crafts and other allied economic activities in rural areas. At the end of June 1985, the percentage of overdues to demand at the PACBs level was around 40 percent, while at the level of LDBs it was around 42 percent. It was worse in the case of RRBs and commercial banks at around 50 percent. The health of agricultural credit institutions, both co-operatives and commercial banks, was in a very poor state in many parts of the country. Wilful default and overdues were mounting even in co-operatively progressive states like Gujarat and Maharashtra.

According to the Seventh Five-Year Plan (1985-90) if this trend was not reversed and if banks were reduced to institutions providing grants rather than recycling credit, the banking system would not be able to meet the credit needs of agriculture in future. The Eighth Five-Year Plan (1992-97) confirmed these apprehensions: The debt relief scheme announced in 1990-91, affected the recovery climate resulting in a lower volume of credit flow. The need for a strong base was recognised. In June 1991, the Government of India announced a New Economic Policy effecting major changes designed to correct the macro economic imbalance and effect structural adjustment so as to bring about a competitive system and promote efficiency in the real sectors of the economy. Hence, in August 1991, the Ministry of Finance appointed a committee on financial systems. The committee suggested, that the entire co-operative credit structure has been built on a weak base, namely the primary credit societies at the village level. Now, if are wishes to take seriously the suggestion of the committee on financial systems (1991), that the concessional credit to the non-viable sector should be gradually phased out, it will amount to gradually winding up co-operative credit institutions at all levels as privileged institutions and brings such of them as will survive at par with the commercial banks.[25]

In the context of lower capability of the co-operative institutions for recycling funds for promoting faster agricultural growth and reduced eligibility for availing of refinance facility from NABARD due to mounting overdues, the Narasimham Committee on Banking sector reforms (1988) suggested de-layering of the co-operative credit system. This is with a view to reducing the intermediation costs and providing the benefit of cheaper NABARD credit to the ultimate borrowers. This will improve the credit delivery system in two ways. First, there will be reduction in the interest rate for the ultimate borrower due to reduction in the level of intermediation. Second, it will enhance flow of refinance in cases where one tier of the structure is not able to borrow on account of their ineligibility due to high percentage of overdues. With the Green Revolution and commercialisation in Indian agriculture, demand for credit in rural areas has increased manifold. Multi-agency approach to rural credit was felt essential to meet the demand and was introduced. Co-operatives, however, remained weak partners not with standing their significant share in credit disbursement. A few initiatives have been taken changing the co-operative structure in the country. Of late, Self-Help Groups (SHGs) are thought to be the answer to the twin problem of restricted availability of credit to weaker sections and non-viability of institutions because of numerous but tiny accounts. SHGs are being linked to the formal system of credit and the linkage programme is very actively pursued by all concerned. These groups are small and homogeneous and are run on principles of co-operation. The SHGs are nothing but limiting cases of PACBs. A few success stories of SHGs reported reassure us that co-operation can succeed even when co-operatives failed. This gives hope that co-operatives can be revived. The rural credit co-operative institutions (RCCIs), one of the strong arms of rural financial institutions (RFIs), have made significant strides in delivering credit for agriculture and rural development. The co-operative movement has taken deep-roots and culminated in a massive rural co-operative institutional structure. Still, co-operatives form the major component of the rural credit system in India. Of the total credit (including short-term, medium-term and long-term) disbursed during the year 1996-97, the share of co-operatives was 49 per cent, about half of the total credit disbursed. The share of the commercial banks and Regional Rural Banks (RRBs), the other two institutional players in the rural credit market, together was 51 percent. In other words, Co-operatives still constitute a dominant segment of the rural credit market.[26]

During the period 1993-94 to 1996-97, while the share of Co-operatives and state governments in direct institutional credit disbursed for agriculture and allied activities declined, that of scheduled commercial banks and RRBs increased. The share of co-operatives in direct institutional credit disbursed for agriculture and allied activities declined steadily from 56.5 per cent in 1993-94 to 49.1 per cent in 1996-97. In accordance with the government's decision to double the flow of credit to agriculture and allied sectors by the end of Ninth Five-Year Plan (1997-2002), the NABARD has been striving towards achieving this target.[27]

The primary agricultural co-operative banks are grass-root level co-operative credit institutions playing a pivotal role in the disbursement of short term rural credit. As at end-March 1998, there were around 92,000 PACBs with a total membership aggregating 997.1 lakhs, of which borrowing membership constituted 41.3 per cent. PACB continued to depend mainly on borrowings from higher financing agencies to fund their operations. At the end of March 1997 the deposits of PACB amounted to Rs.5,255 crores, while the borrowings stood at Rs.13,299 crores. The loans issued amounted to Rs.12,594 crores, while the loans outstanding amounted to Rs.15,323 crores.[28]

Table 3.2

Primary Agricultural Co-operative Banks-at a Glance in India Level

(Rupees in Crore)

Sl. No.	*Particulars* 1	*1994-95* 2	*1995-96* 3	*1997-98* 4
1.	Total number of PACBs	95,946	93,314	92,000
2.	Membership (000's)	1,21,802	94,420	99,710
3.	Owned funds	6,279	3,720	NA
4.	Deposits	19,825	3,594	NA
5.	Working capital	40,967	20,846	NA
6.	Loan advances	26,420	10,883	16,081
7.	Outstanding	24,559	12,979	18,175
8.	Overdues	5,357	4,560	NA

Source: Statistical Statement Relating to the Co-operative Movement in India, 1995-96, Part-I Credit Societies NABARD, p.15.

There were around 90,245 Primary Agricultural Co-operative Banks (PACB) operating in the country as of end-March 1999, with the total membership of these PACB aggregating 1,016 lakhs, and borrowing members at 440 lakhs constituting around 43 per cent. The high level of overdues in many states has made a large number of members ineligible for fresh borrowings. As at end-March 1998, deposits of PACS amounted to Rs.6,518 crores representing a growth of 15 per cent over the previous year. The outstanding amount of borrowings of all PACB stood at Rs.17,073 crores as at the end of March 1998. The loans issued amounted to Rs.16,081 crores, while the loans outstanding amounted to Rs.18,175 crores at the end of March 1998.[29]

The discussion reveals that the co-operative movement has been brought to India from other countries. Though pre-independence era seen the co-operative credit movement, the independence and five-year plans made the movement to grow faster and diversify its activities in many sectors.

REFERENCES

1. Mathur, B.S. *Co-operation in India,* Sahitya Bhawan, Agra, 1988, p. 44.
2. Kulkarni, K.R., *Theory and Practice of Co-operation in India and Abroad,* Vol.II (Part II), Co-operators Book Depot, Bombay, 1965.
3. Bedi, R.D. *Theory, History and Practice of Co-operation,* Loyal Book Depot, Meerut, 1986-87, p.55.
4. Maslennikov, V., *The Co-operative Movement in Asia and Africa,* Progress Publishers, Moscow, 1983, p. 8.
5. Mamoria, C.B. and SAKSENA.R.D., *Co-operation in India,* Kitab Mahal, Allahabad, 1977, p. 87.
6. Bansil, P.C., *Agricultural Problems of India,* Oxford and IBH Publishing Co, New Delhi, 1981, p. 207.
7. *Co-operation in India and Abroad,* Tej Bahadur Singh Chandel, Kanpur, 1968, p. 1.
8. Kulkarni, K.R., *Theory and Practice of Co-operation in India and Abroad,* Vol.I. Co-operators Book Depot, Bombay, 1968, p. 2,5.
9. Mathur, B.S., *Co-operation in India,* Sahitya Bhawan, Agra, 1984, p. 61,62,65.
10. Dandekar, M.M., *The Indian Economy 1947-1992, Agriculture,* Vol.I, Sage Publication, India Pvt. Ltd., New Delhi, 1994, p. 340.

11. Bhatia. B.S., Verma. H.L., and Mahesh C. Garg, *Encyclopaedia of Co-operative Management, Vol.3, Co-operative Banking: Levers of Rural Economy*, Deep and Deep Publications, New Delhi, 1994, pp. 210-212.
12. Nakkiran, S., *Co-operative Banking in India*, Rainbow Publications, Coimbatore, 1980, pp.1-5.
13. Op.cit. Mathur. B.S. p. 65.
14. Op.cit. Bhatia. B.S., Verma. H.L., and Mahesh C. Garg, p. 120.
15. Eleanor M. Hough, *The Co-operative Movement in India*, Oxford University Press, 1966, pp.44-47.
16. Mamoria, C.B., *Agricultural Problems of India*, Kitab Mahal Publication, Allahabad 1995, p. 526, 527.
17. Op.cit. Mathur, B.S. p. 68.
18. Sami Uddin and Mahfoozur Rahman, *Co-operative Sector in India: After Independence*, S.Chand and Company Ltd., New Delhi, 1983, p. 28.
19. Ibid. p. 28,29.
20. Ibid. p. 29,30.
21. Tandon, B.B and Kulwinder Kaur, *Indian Economic Problems*, S. Chand and Company (Pvt) Ltd., New Delhi, 1987, p. 185.
22. Jain, P.C., *Economic Problems of India*, Chaitanya Publishing House Allahabad, 1965, p. 148.
23. Op.cit. Sami Uddin and Mahfoozur Rahman, p. 33, 47.
24. Op.cit. Mamoria.C.B. p. 549.
25. Op.cit. Dandekar, M.M. p. 356, 367.
26. Satyasai, K.J.S. and Badatya.k, C., "Restructuring Rural Credit Co-operative Institutions", *Economic and Political Weekly*, Jan 29, Feb 4, 2000, Vol. XXXV, No. 5, p. 328,323,307.
27. Report on currency and Finance (1997-98) Vol.I, Economic Review, *Reserve Bank of India (RBI)*, Mumbai, P.V13, V15, V16.
28. Report on Trend and Progress of Banking in India 1998-99, Supplement to *RBI Bulletin*, December 1999, Mumbai, p.117.
29. Report on Trend and Progress of Banking in India 1999-2000. Supplement to *RBI Bulletin*, December, 2000, p.106.

Chapter - 4

PROFILE OF THE STUDY AREA

INTRODUCTION

This chapter gives a descriptive account of the area of study and gives the general features such as agricultural growth, patterns of cultivation, occupational distribution, land use patterns, irrigation, institutional role in the district and other characteristics pertaining to the district level and two villages where two sample PACBs are located. An analysis of the brief history of the study area, social status, size and growth of population, literacy level, infrastructure facilities, and administrative characteristic would provide a good backdrop for the results and discussions of the study and to draw meaningful conclusions.

LOCATION AND AREA

Vellore district, carved out of North Arcot district, came into being on 30 September 1989. In the 19th century the district lay between 12°21' and 14°10' north latitude and between 78°14' and 80°13' east longitude[1]. Present geographical area of the district is 6077 sq.kms and the district lies between 12°15' to 13°15' North latitudes and 78°20' to 79°50' East longitudes in Tamil Nadu State[2]. On the west are the districts of Dharmapuri and Chittoor (Andhra Pradesh). Kanchipuram and Thiruvannamalai districts are its eastern and western boundaries. The district is an inland region and is of irregular shape. Headquarters of this district is Vellore. The climate of the district is basically tropical. The average minimum and maximum temperatures in the district normally ranges between 16°C and 38°C touching on a few summer days more than 40°C.

Fig. 4.1: A Map of Tamil Nadu

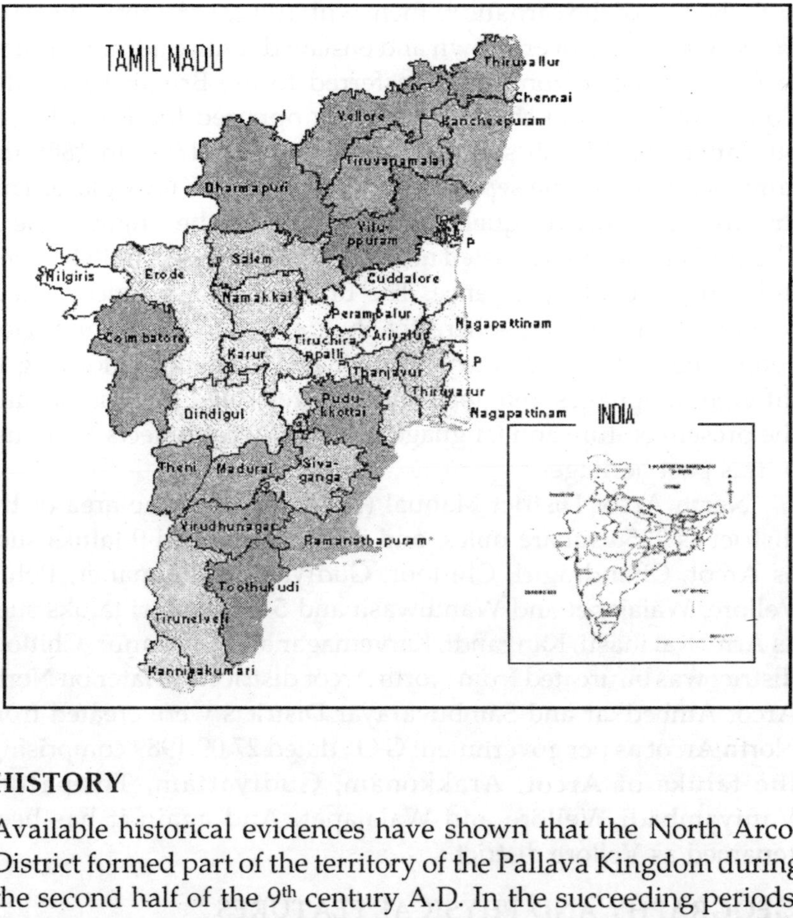

HISTORY

Available historical evidences have shown that the North Arcot District formed part of the territory of the Pallava Kingdom during the second half of the 9th century A.D. In the succeeding periods, the district experienced many trails and tribulations under the rule of several dynasties like the Cholas, Rashtrakutas, the Thanjure Ruler, the Hindu Ruler of Vijaya Nagar and the Muslim Ruler of Deccan. During the 18th century several battles have taken place in the district. In the war on the coromandel coast, the English captured Arcot, and this is considered as a memorable event in Anglo-Indian history. Thereafter North Arcot becomes the theatre for many battles for supremacy among British, French and Karnatic rulers.

In 1787, the Nawab had assigned the revenues of the Karnatic to the East India Company and North Arcot then came under their management. In 1801 the Nawab Azim-ud-Daula ceded region with the rest of the carnatic to them with full sovereignty. By 1801, the Nawabs were overthrown and enslaved. Entire administration of the carnatic region was transferred to the British East India Company. The local chieftains initially opposed the English rule but latter the English subdued all of them by 1805. In 1806 the famous mutiny of the sepoys stationed at Vellore took place. This mutiny was however quelled successfully by the English rulers. The British rule, which lasted nearly 150 years, ended in 1947 when India attained independence. The British had introduced many socio-economic changes here for the development of the region before they left this country. Thus many dynasties belonging to different languages, religions and cultures ruled this district and the present culture and language of North Arcot reflects a mixture of this past heritage.

North Arcot District Manual (1894) reported the area of the district as 7256 square miles[3] and it then contained 9 taluks such as Arcot, Chandragiri, Chittoor, Gudiyattam, Palmaner, Polur, Vellore, Walajapet and Wandiwash and 5 Zamindari taluks such as Arni, Kalahasti, Kangandi, Karvetnagar and Punganur. Chittoor district was bifurcated from North Arcot district and later on North Arcot Ambedkar and Sambuvarayar Districts were created from North Arcot as per government G.O., dated 27.09.1989 comprising, the taluks of Arcot, Arakkonam, Gudiyattam, Tirupattur, Vaniyambadi, Vellore and Walajapet. And again it has been renamed as Vellore district.

GEOGRAPHY AND PHYSICAL FEATURES

The district has in its boundary sizable hilly and mountainous regions. The mountains of the Eastern Ghats and their spurs extend over the north-western region of the district and they add to the scenic beauty of the district. The slope is from west to east and the land in the eastern part is flat and level. The soil is fertile and mostly well utilised for cultivation. There are a number of tanks/lakes and the rocky hills enable the formation of such tanks and lakes through streams in view of their relatively poor and free flow of rain waters. A number of big tanks, thus formed, exist in the district. The tank at Kaveripakkam, for example, extends to about

20 square kilometers of area. The Javadi hills represents the important mountain range in this district and they are lofty mountains. The Javadis cover the eastern parts of Tirupathur and Vaniyambadi taluks with spurs extending into Vellore taluk, upto about 10 kilometers from the city of Vellore. The general height of the Javadi hills is 2500 feet with peaks raising upto 4200 feet. The Elagiri hills lie in the north-eastern part of Tirupathur taluk. Besides the above, there are many hills (of low elevation) which lie scattered in Vellore, Vaniyambadi, Gudiyatham, Arcot and Walajapet taluks.

The forests in Gudiyatham taluk are of fuel and scrub type. The reserved forests in the other taluks include of hill forests, plain forests, scrub growth and Palar river plantations. The hill forests contain sandalwood and bamboo. The estate forests taken over in Tirupathur taluk contain scrub jungle. Forest improvement schemes have been undertaken during the Five year plans and as a consequences of the efforts in this regard, the area under forest has also been gradually increasing in this district. During the plans, the acreage under fuel and fodder species in forests has improved. The forest regions of Vellore and Tirupathur fall under the Vellore circle and the area of reserve-forests and reserve lands here (as in 1990-91) are 305624 hectares and 9938 hectares respectively. The chief forest produce here are sandalwood, other fuel wood and bamboo.

RIVERS, CANALS AND WATER WAY

The district has the following important rivers: Palar, South Pennar, Cheyyar and Ponnai. Barring a few days in the rainy season, these rivers are mostly dry and are sandy wastes. 'Palar' is the most important river of the district. Raising near Nandidurg in Karnataka state, Palar enters the district in the western part through Vaniyambadi taluk. It flows then in an east-north-easterly direction in the taluk and is joined by some jungle streams like Malattar flowing in Gudiyattam taluk. The river forms the boundary between Vaniyambadi and Gudiyattam taluks for some distance, then strikes an eastern course and runs along the boundary between Gudiyattam and Vellore taluks. After cutting across Walajapet taluk, it runs along the boundary between

Walajapet and Arcot taluk as also Arakkonam and Cheyyar taluks before entering Chengalpattu district on the east.

Cheyyar river is the next important river of the district. It rises in Javadi hills, flows in a southern direction at first and turns south-eastwards. Ponnai river which is a tributary of Palar river rises in Chittoor district of Andhra Pradesh. It flows on the western part of Walajapet taluk and joins Palar near Ranipettai. The rivers, Goddar and Kavundinya Nadi which flow from north to south in Gudiyattam taluk are also tributaries of the Palar river. They join the Palar near Ambur and Gudiyattam respectively.

CLIMATE AND RAINFALL

The district has a moderate climate. Some taluks like Arakkonam, parts of Walajapet and Gudiyattam that are interspersed by hills and mountains are exposed to different climatic conditions and having very hot weather during the winter seasons. In Tirupathur and Vaniyambadi taluks, climate is rather slightly cold during winter, but moderate in all other seasons.

The district receives rainfall due to southwest and northeast monsoons. The first rainy season occurs between June and September, the second one during the retreating monsoon period of October and November. The region gets heavy rainfall during the southwest monsoon period of June to September. The normal total annual rainfall of the district is 935.4 mm. This is slightly more than state average figure of 923.1 mm. Normal rainfall in different months are 448.8 mm during June to September, 379.3 mm during October to December, 23.2 mm during January to February and 102.1 mm during March to May. The district has a long dry season from January to June and has an average of 100 to 150 dry days per annum. Tanks and wells are the important sources of irrigation in Vellore district and most of them depend on rainfall for the supply. Inadequate rainfall in some years is the great drawback of the district and hence whenever famine hit the state, the composite North Arcot was not an exception[4].

Cyclonic storms are common, in May and October months at the change of the monsoon. These cyclones sometimes cause heavy rain. During April to June, it is generally very hot in most parts of the district and the temperature rises to a maximum of exceeding

40°C, in Vellore and its surroundings. Mean monthly temperature is 30°C from March to June. Some parts like Tiruppathur and Elagiri hill regions have moderate temperatures in summer. High evaporation in Vellore renders rainfall less available for crops and for the recharging of underground water.

SOIL

Soil of the district is not uniform and one can observe different varieties of soil. While the central and southern parts of the district are mostly hilly, the eastern portion is almost a stretch of unbroken plain. In the western sector the land rises gradually towards the Mysore plateau along the Kangundi Zamindari of Chittoor district. The soil is mostly of the red ferruginous variety both sandy and loamy, with black area accounting for about 16 per cent. The black soil is found mostly in the neighbourhood of the rivers of Palar, Ponnai and in the anaicuts of a few big tanks. The black series clay occurs in Gudiyattam, Walajapet, Arcot, Arakkonam and Vellore taluks. The black series loam is found mostly in Arakkonam and Gudiyattam taluks in larger areas than in other taluks while red series loam is found in all the taluks.

FLORA AND FAUNA

Some varieties of flora and fauna, most of which are unique to the region are found in Javadi region. On flora, no distinguishing variety is observed in the district. The hilly areas belong to the 'dry deciduous' type generally observed in the projections of the Eastern Ghat ranges. In the lower areas, plants like tamarind trees, teak wood, sandal wood, ven-teak, casurina, bamboo etc., can be found. With regard to fauna, no special and distinguishing variety is available in the district. But the western region, particularly the mountainous tracts possess some wild animals. Bison is seen in small numbers on the Javadi ranges and tiger is seen occasionally. Leopards are found all through the rocky hills. Black bears, hyenas, sambar, spotted deer, jungle sheep, deer, antelope, and different varieties of snakes are spread over different regions of the district.

DEMOGRAPHIC PROFILE

Table 4.1 gives a clear picture of the Demographic profile of the district.

Table 4.1
Demographic Profile

Particulars	*Tamil Nadu*			*Vellore District*		
	1980-81	*1990-91*	*2000-01*	*1980-81**	*1990-91*	*2000-01*
a. Persons	48,4080,77	55,858,946	62,110,839	4,414,324	3,026,432	3,482,970
Male	24,487,624	28,298,975	31,268,654	2,230,559	15,29,944	1,743,871
Female	23,920,453	27,559,971	30,842,185	2,183,765	1,496,488	1,739,099
b. Decadal growth rate						
	1971-1981		17.50			17.53
	1981-1991		15.39			15.14
	1991-2001		11.19			15.09

Source: Census of India 1981, 1991, 2001. Website: www.tngov.in
* Composite

Census 2001 result has shown that Vellore district has a male population of 17,43,871 and female population of 17,39,099 which totals to 34,82,970. It is the third most populous district of the state. Decadal growth rate of population of the district has been reduced from 15.14 in the period between 1981-1991 to 15.09 for the period 1991-2001. At the same time decadal growth rate of the state as a whole has been reduced from 15.39 in 1981-1991 to 11.19 in 1991-2001. Thus the decadal growth rate of the district is much higher compared to that of state.

SEX RATIO AND DENSITY OF POPULATION

Table 4.2 clearly shows sex ratio and density of population in the district.

Table 4.2
Sex ratio and Density of Population

Particular	*Sex ratio*			*Density of population*		
	1981	*1991*	*2001*	*1981*	*1991*	*2001*
Tamil Nadu	977	974	986	372	429	478
Vellore district	979	978	997	360	498	573

Source: Census of India, 1981, 1991, 2001. Website: www.tngov.in.

Sex ratio has increased from 979 in 1981 (undivided) to 997 in 2001. It is higher than the state figure of 986 in 2001.

Density of population in district has increased highly as it was 360 in 1981 census (undivided) and 573 in 2001 census. It is higher than the state figure of 478 in 2001 census. Thus the present Vellore District seems to be thickly populated and this may be because of more number of municipal towns/urban pockets in this districts.

LITERACY LEVEL

The average literacy rate in the district was 40.89 per cent as 1981 Male literacy rate was 53.48 per cent where as that of female was 28.04 per cent. The district average among male and female stood at 82.67 per cent and 63.53 per cent in 2001. The average literacy rate in the district was 73.07 per cent as per 2001 census. It is show us Table 4.3.

Table 4.3
Literacy Rate

(in percentage)

Particulars	1981	1991	2001
Tamil Nadu			
Persons	46.76	62.66	73.47
Male	58.26	73.75	82.33
Female	34.99	51.33	64.55
Vellore District			
Persons	40.89	60.87	73.07
Male	53.48	72.94	82.67
Female	28.04	48.58	63.53

Source: Census of India, 1981, 1991, 2001, District Statistical Hand Book.

The district which was away from the state average literacy rate in 1981 has come very close to the state average in 2001.

OCCUPATIONAL STRUCTURE OF THE DISTRICT

The occupational structure of the district reveals the economic activities of individuals in various sectors of the economy. It can be observed from Table 4.4 that Vellore district is predominantly agriculture oriented and a greater proportion of the economically active population is involved in agriculture and allied activities. The industrial units have come up at Ranipet and an a few other places in recent years, unorganised sector continues to contribute a greater proportion besides the traditional handloom weaving sector, and leather tanneries. Vellore district is one of the largest beedi making centres in India. Recently there is a recession not only in leather tanneries but also in the beedi industry. Mat weaving industries are also functioning in the villages as cottage industries in a few locations.

Table 4.4
Distribution of Workforce in Vellore District

(in percentage)

Sl. No.	*Workforce*	*1961*	*1971*	*1981*	*1991*	*2001*	*Change (1961-2001)*
1.	Cultivators	51.44	38.27	35.29	23.40	25.48	-25.96
2.	Agricultural labourers	17.07	31.22	32.53	32.99	35.91	18.84
3.	Household Industry	06.95	04.10	06.04	06.98	11.87	4.92
4.	Other workers	24.54	26.41	26.14	36.63	26.74	2.2

Source: Computed from primary census abstract 1961, 1971, 1981 and 1991. Annual credit plan 2000-2001, Vellore District.

Table 4.4 reports the distribution of the workforce in the district according to the 1961, 1971, 1981, 1991 and 2001 population census. In 1961 census 51.44 per cent of the population were cultivators. It has come down to 25.48 per cent of the population by 2001 census. It has been reduced by 28.04 per cent point over four decades. No other category has shown a change to this magnitude. Agriculture labour has increased its share by 18.84 per cent and it is 35.91 per cent to the total workforce by 2001 census. In 1961 census household industry was 6.95 per cent and it has increased by 4.92 per cent during 40 years. The comparison is valid only upto a point since the conceptual definition for work force and category of workers differ from one census to another.

LAND UTILISATION PATTERN

It is evident from Table 4.5, that total geographical area is 592018 hectares. It is 4.56 percent of the total geographical area of the state. Area covered by forest is more in the district, i.e., 1,61,229 hectares or 27.23 percent of the total geographical area of the district and it is 7.56 percent of the total forest in the state land put to non-agricultural use are 82,685 and 83,069 hectares in 1998-99 and 2000-2001. The net area sown at 2,48,279 (41.94 percent) in 1998-99 and it was declined by 2,10,956 (35.63 percent) hectares in 2000-2001 respectively.

Table 4.5
Land use Pattern in Vellore District During 1998-1999 to 2000-2001

(Area in hectares)

Classification	*1998-1999*	*1999-2000*	*2000-2001*	*Percentage share to state (as per 1999-2000)*
Forest	1,61,229 (27.23)	1,61,229 (27.23)	1,61,229 (27.23)	7.56
Barren and uncultivable land	26,054 (4.40)	26,054 (4.40)	26,054 (4.40)	5.48
Land put to non-agricultural use	82,685 (13.97)	82,720 (13.97)	83,069 (14.03)	4.18
Cultivable waste	6,709 (1.14)	6,757 (1.14)	6,757 (1.14)	1.94
Permanent pastures and other grassing land	3,833 (0.66)	3,833 (0.65)	3,833 (0.66)	3.13
Land under miscellaneous tree crops and groves not included in net area sown	1,830 (0.31)	2,685 (0.45)	3,116 (0.53)	1.10
Current fallows	30,574 (5.17)	60,997 (10.30)	86,974 (14.69)	5.62
Other fallow land	30,825 (5.21)	31,827 (5.38)	10,030 (1.69)	2.79
Net area sown	2,48,279 (41.94)	2,15,916 (36,47)	2,10,956 (35.63)	3.95
Geographical area	5,92,018 (100.00)	5,92,018 (100.00)	5,92,018 (100.00)	4.56

Source: Statistical Hand Book-2000, Government of Tamil Nadu.
Annual credit plan Vellore District 2000-2001.

LAND HOLDING STRUCTURE

The pattern of distribution of the size of holdings in the district (1997-98) is shown below.

Table 4.6 gives a clear picture of the operational land holdings of the district. One can observe that there are 4,23,771 households having operational land holdings and it accounts an area of 2,86,971.33 hectares, of which 2,46,690 (58.21 per cent) have just less than 0.5 acres of land and it accounts 18.98 per cent of total area. High number of small land holdings in this region has been observed by many authors (John Harris, 1991). Considering the

Table 4.6
Area of Operational Land Holding (1997-98)

Sl.No.	Land size	Number	Percentage	Area in Hectare	Percentage
1.	< .5	2,46,690	58.21	54,477.61	18.98
2.	.5 - 1	92,570	21.85	65,419.32	22.80
3.	1 - 2	58,242	13.74	79,225.43	27.61
4.	2 - 3	15,351	3.62	36,402.72	12.69
5.	3 - 4	5,446	1.29	18,508.21	6.45
6.	4 - 5	2,508	0.59	10,802.63	3.76
7.	5 - 7.5	2,050	0.48	12,190.39	4.25
8.	7.5 - 10	542	0.13	4,584.20	1.60
9.	10 - 20	336	0.08	4,177.06	1.45
10.	Above 20	36	0.61	1,183.77	0.41
	Total	4,23,771	100.00	2,86,971.33	100.00

Source: Statistical Hand Book 2000-2001.
Statistics Department, Vellore District.

number of land holdings having less than 2 acres of land which accounts for 93.81 per cent of total land holdings and 69.39 percentage of total area, one can easily say that this district has more number of marginal holdings which cannot assure adequate income to the family. There are only 8.78 per cent land holdings having more than 10 acres of land and it accounts for 1.86 per cent of the total area. There are no big estates or Zamindaris in the district.

AGRICULTURE

Agriculture still occupies the dominant place in the economy of Vellore district. Among the major crops of the district, paddy, groundnuts and sugarcane are predominant. These crops also support agro-processing industry that is an important part of the manufacturing sector. Banana, coconut and mango are also grown in recent times.

Agriculture in the district is depending on rainfall pattern. Paddy has been traditionally grown here in well-defined seasons, namely *Samba, Navarai and Sornavari*. The *samba* crop is the main rainy season crop. It is sown in July or August and harvested in December (or) January. The *navarai* crop coincides with the dry season and depends entirely on irrigation. It stretches from December (or) January to May. The *swarnavari* crop extends from

May (or) June to September and encompasses the light, southwest monsoon. Millets and sorghum are grown as crops from June - July to October - November, and as irrigated crops from February - March to June - July. For groundnuts, the rain fed season is from June - July to September - October, and the irrigated crop is grown between December - January and March - April. The main cropping season for pulses (red gram, black gram and green gram) is from June - July to December - January. Sugarcane and bananas are planted in January and harvested in December.

According to the North Arcot district manual, 1894 prepared by Arthur F. Cox, agricultural labourers were relatively few in the district. The ratio was 2 labourers to 11 farmers, and this low proportion shows the great extent to which peasant proprietorship prevailed. About one quarter of the agricultural labourers were farm servants who were employed throughout the year; the rest were engaged for special work only, such as transplanting and harvesting (or) if employed longer, serve for monthly wages and can be dispensed with when work is slack. Some of the general labourers are also employed on agricultural activities when demand arises.

Large number of small farmers in the district is one thing that requires special mention. In 1979 there were 5,74,000 holdings and average size was 1 hectare. About 68 per cent farmers were operating 1 hectare (or) less, and about 86 per cent were operating 2 hectares (or) less. Owner-operated farms dominated the district. Pure tenant farmers are scarce and most land-leasing arrangements involve farmers who already own some land of their own. Rents are paid in cash (or) kind, but they usually involve fixed rents. Share cropping is rare[5]. During 1999-2000, 80 per cent of the land holdings are of size less than 1 hectare, and 14 per cent are of size between 1 and 2 hectares. While 5 per cent of the land holdings fall in the category 2 to 5 hectares, only 1 per cent of the holdings are of size above 5 hectares[6].

One can observe from the Table 4.7 that groundnut and paddy are important crops in the district. The district's share in the total area under groundnut cultivation in the state varies between 6.97 and 9.45 during the period discussed below.

The Table 4.8 clearly points out the trend of the production of selected crops. The data indicate that paddy production has increased nearly three times. Production of sugarcane and

Table 4.7

Area of Selected Crops in Vellore District During 1990-1991 to 1998-1999

(in hectares)

Year	*Paddy*	*Cholam*	*Cumbu*	*Ragi*	*Sugarcane*	*Cotton*	*Groundnut*
1990-91	34567	24804	4629	8352	20814	2663	74109
	(1.86)	(4.58)	(1.69)	(4.91)	(8.39)	(1.11)	(7.69)
1991-92	70437	25311	3647	8754	19753	3925	86078
	(3.32)	(4.94)	(1.48)	(5.54)	(8.29)	(1.52)	(7.83)
1992-93	48502	18215	4615	9984	18011	7123	82917
	(2.22)	(3.76)	(2.09)	(6.61)	(8.33)	(2.66)	(6.97)
1993-94	61963	17924	3693	8815	21961	6340	91538
	(2.68)	(3.54)	(1.73)	(5.59)	(8.81)	(2.76)	(7.90)
1994-95	52681	1605	2237	8046	27596	9283	86063
	(2.36)	(0.37)	(1.16)	(5.54)	(8.41)	(3.364)	(7.96)
1995-96	60524	13943	2108	7317	29187	16717	76556
	(3.10)	(3.64)	(1.21)	(5.99)	(8.95)	(6.40)	(8.20)
1996-97	86846	11860	1893	7390	25339	18011	70248
	(3.99)	(3.00)	(1.14)	(6.65)	(9.74)	(7.14)	(7.78)
1997-98	71607	8963	1511	5956	29072	10385	62236
	(3.16)	(2.35)	(0.89)	(5.56)	(10.27)	(4.55)	(7.17)
1998-99	77129	13680	4023	10657	26646	7284	81147
	(3.39)	(3.74)	(2.61)	(8.88)	(8.70)	(3.32)	(9.45)

Figures in the brackets are respective percentage share in the state.
Source: Statistical Hand Book - 2000, Government of Tamil Nadu.

Table 4.8

Production of Selected Crops in Vellore District During the Period 1990-1991 to 1998-1999

(in tonnes)

Year	*Paddy*	*Cholam*	*Cumbu*	*Ragi*	*Sugarcane*	*Cotton*	*Groundnut*
1990-91	100850	29030	5740	17170	1598640	3720	104360
	(1.74)	(5.29)	(1.94)	(5.43)	(6.81)	(0.91)	(8.85)
1991-92	223100	19210	7050	18890	1322560	5920	151410
	(3.38)	(3.47)	(2.59)	(6.07)	(5.31)	(1.44)	(9.97)
1992-93	107950	23980	6680	26900	1517640	10180	135270
	(1.59)	(4.93)	(2.66)	(9.24)	(6.58)	(2.24)	(7.66)
1993-94	186690	28760	5550	23400	1523810	10170	185710
	(2.77)	(5.92)	(2.33)	(7.07)	(5.86)	(2.39)	(9.95)
1994-95	146060	22360	3890	17300	2247920	16840	201360
	(1.93)	(4.56)	(1.68)	(6.07)	(6.17)	(3.83)	(11.43)
1995-96	168730	18360	3540	14640	2609290	30150	145820
	(3.19)	(5.56)	(1.87)	(6.62)	(7.96)	(8.88)	(9.59)
1996-97	260570	16980	3310	15680	2100020	22340	129740
	(4.49)	(4.69)	(1.70)	(8.21)	(8.10)	(6.77)	(9.02)
1997-98	236910	8030	2080	12360	2550840	12640	109770
	(3.44)	(2.12)	(0.98)	(5.67)	(8.45)	(3.53)	(7.80)
1998-99	295870	16930	7000	30930	251090	7580	148900
	(3.63)	(4.59)	(3.40)	(12.85)	(7.44)	(1.87)	(9.49)

Figures in the brackets are respective percentage share in the state.
Source: Statistical Hand Book - 2000, Government of Tamil Nadu.

groundnut also have increased. There seems to be an increase in the production of all crops except in case of cholam.

A comparative position of the productivity of various crops grown in the district vis-a-vis that of the state and nation is furnished in the Table 4.9 given below.

Table 4.9
Productivity of Various Crops During 2000-2001

(Tons/hec)

Sl.No.	*Crops*	*Productivity*		
		District level	*State level*	*National level*
1.	Paddy	3.31	3.05	1.89
2.	Oil seeds	0.89	0.77	0.84
3.	Coarse cereals	1.15	1.11	1.77
4.	Sugarcane	89.00	108.60	70.00
5.	Pulses	0.33	0.45	0.57
6.	Cotton	0.21	0.27	0.21

Source: Potential linked credit plan - 2001-2002 Vellore District, NABARD.

Data indicate that much has to be done to increase the productivity of sugarcane, pulses and cotton. A significant effort has to be taken in sugarcane and purses.

IRRIGATION

Agricultural activity in the district is entirely dependent on the seasonal rainfall, canals, tanks, wells, springs and channels. The Anaicut across the rivers Palar, Ponnai and Pennaiyar meet the major part of the irrigation needs in the past.

Table 4.10 shows that net area irrigated in Vellore district is 125275 hectares in 1998-99 and against 98618 hectares in 1993-94 while wells (other than tube) in the district forms a very important source of irrigation covering more than 80 per cent of the total area under irrigation. This is followed by tanks with 22.14 per cent, and just 1.61 per cent by canals and negligible area by other sources like springs and channels. One can observe from the table that 7.6 per cent of the net area irrigated by wells (other than tube) and 4.02 per cent of the area irrigated by tanks in Tamil Nadu as a whole is in this district. Since wells and tanks need adequate rain to replenish water recharge, they provide only limited insurance against drought. This is particularly trouble some because the region experiences wide variations in annual rainfall.

Table 4.10
Source-wise Net Area Irrigated in Vellore District During 1993-94 and 1998-99

Sl.No.	*Source*	*1993-94*	*1998-99*	*Percentage Share to State (1998-99)*
1.	Government canals	2913 (2.95)	2020 (1.61)	0.24
2.	Private canals	-	0 (0)	0
3.	Tanks	15074 (15.29)	27739 (22.14)	4.02
4.	Tube wells	-	223 (00.18)	0.10
5.	Other wells	80538 (81.67)	95153 (75.96	7.60
6.	Other sources	93 (0.09)	140 (0.11)	4.15
	Total	98618 (100.00)	125275 (100.00)	4.15

Source: Statistical Hand Book - 2000, Government of Tamil Nadu. Annual statistical abstract of Tamil Nadu 1993-94, Chennai.

There are 119699 wells in the district, of which 14047 are fitted with oil engines, 112170 with electric motors and 9050 with country mhots during 2000-2001.

ANIMAL HUSBANDRY

Table 4.11 shows the results of livestock census. District has 6.46 per cent of state's sheep population and 6.36 per cent of state's cattle population but has only 4.20 per cent of poultry, 3.77 per cent of buffaloes, 3.26 per cent of goats, and 3.24 per cent of pigs population. Thus the district is better in livestock position which can enhance their income from dairy farming.

Table 4.11
Livestock Census in 1997

	Cattle	*Buffaloes*	*Sheep*	*Goats*	*Pigs*	*Poultry*
Vellore district	575697 (6.36)	103258 (3.77)	339480 (6.46)	209396 (3.26)	19712 (3.24)	779059 (4.20)
Tamil Nadu	9046538	2741263	5258884	6416204	609176	18547825

Figures in the brackets are respective percentage share in the state.
Source: Statistical Hand Book - 2000, Government of Tamil Nadu.

Table 4.12
Livestock Census in the District Given below During the Period 1997 and 1999

Sl. No.	Name of the livestock	1997	1999	Change (1997-99)
1.	Cattle	575697 (28.41)	349203 (21.46)	-6.95
2.	Buffaloes	103258 (5.09)	100419 (6.17)	1.08
3.	Sheep	339480 (16.75)	317194 (19.49)	2.74
4.	Goats	209396 (10.34)	168769 (10.37)	0.03
5.	Pigs	19712 (0.97)	21641 (1.33)	0.36
6.	Poultry and Fowls	779059 (38.44)	670335 (41.18)	2.74

Figures in the brackets are percentage to total.
Source: (a) Statistical Hand Book - 2000, Government of Tamil Nadu.
(b) Annual Credit Plan 2000-2001 Vellore District, Vellore, p.10.

The district had cattle population of 28.41 per cent in the year 1997 and it has come down to 21.46 per cent by 1999. It has been reduced by 6.95 per cent. No other category has shown a change to this magnitude. The total number of goats and pigs have experienced a marginal increase of 0.03 and 0.36 per cent respectively.

INDUSTRY

The Vellore district is fast developing into an important industrial region of the state. There are 919 large and medium scale factories in this district concentrating on the following lines of industrial activity: fertilizer production, porcelain making, manufacture of food products, steel casting, synthetic fibres, chemicals, textile products, auto bearing manufacturing.

'Beedi' is produced on a large scale in this district and this is the second important industry from the point of view of employment of workers. Leather industry, cotton weaving, match industries are some of the major industries of the district. Match industry is one of the predominant cottage and small-scale industries of the district and of the latest origin. It provides employment to the farmers in the off-season and particularly for women and children. There seems to be a high possibility of presence of child labour in this district because of match industry, beedi industry and weaving industry.

According to Arthur F. Cox during the 19th century, weaving was the most important occupation of the people of the district next to agriculture.

According to the census 1991 there were about 5590 small-scale industries in Vellore district employing about 20309 people. End of March 1998 there 13708 were small-scale industrial units employing 67,724 in Vellore districts.

Table No. 4.13 gives information about the Annual Survey of Industries in Tamil Nadu. There were 19177 factories in 1994-95 and which has been increased to 20478 in 1998-1999. The casual workers were 933430 during the year 1994-95 and they have been increased to 961286 in 1998-1999. The number of employees have been reduced from 1374588 to 1168550 between 1994-94 to 1998-99. The total value of output has increased by 1624368 during the year 1994-95 to 1998-99.

District Industries Centre is the implementing arm of the Centre and State governments of various schemes and programmes. District industries centre is an integral part of the Department of Industries and Commerce functioning at the district level.

Table 4.13
Survey of Industries during 1994-95 to 1998-99 in Tamil Nadu

Sl. No.	*Characteristics (1)*	*Unit (2)*	*1994-95 (3)*	*1995-96 (4)*	*1996-97 (5)*	*1997-98 (6)*	*1998-99 (7)*
1.	Factories	Number	19177	19301	19202	19406	20478
2.	Productive Capital	Rs. Lakhs	3669093	4152330	4672144	4992438	4412012
3.	Workers	Number	933430	962935	971435	1064393	961286
4.	Wages to Workers	Rs. Lakhs	239496	282438	320674	385088	296554
5.	Employees	Number	1374588	1210924	1266282	1347669	1168550
6.	Emoluments to Employees	Rs. Lakhs	393998	454591	518605	606960	579873
7.	Total Input	Rs. Lakhs	4436389	5220026	5755994	7591999	5737575
8.	Total Output	Rs. Lakhs	5775188	6747836	7421247	8965363	7399556
9.	Net Value Added	Rs. Lakhs	1156167	1320547	1348652	1036324	1232076

Source: Tamil Nadu—An Economic Approvisal 2001-2002.

There are 15738 registered SSI units in Vellore district, which give employment to 50805 persons and there are 19 medium and large industries in the district in which 9590 persons are working. There are 11 industrial co-operative societies in Vellore district[7].

Under PMRY, 837 persons benefitted in the year 1998-99 and the received an amount of Rs.362.40 lakhs and in 1999-2000, numbers of beneficiaries were 534 and they received an amount of Rs.236.63 lakhs. Under women entrepreneurs development training programme 3527 females have been trained in the year 1999-2000.

FINANCIAL INSTITUTIONS

The district is served by 203 branches of commercial banks besides branches of DCCB Ltd. TNSARDB, Urban Co-operative Banks and TIIC.

Table 4.14 shows the particulars of financial institutions in Vellore district by the end of March 2000. There were 29 banks working in the district and among them 18 are nationalised banks, 11 private banks and 1 District Central Co-operative Bank with 35 branches over the district and 185 PACBs were functioning in the district.

All these banks have a good number of branches throughout the district among that 103 were working in rural areas 66 were working in semi-urban areas and 34 in urban areas.

Table 4.14
Particulars of Financial Institutions in Vellore District

Type	*Nos.*	*No. of Branches as on March 2002*			
		Rural	*Semi-Urban*	*Urban*	*Total*
State Bank and Associates	4	19	13	6	38
Public Sector Banks	14	69	43	17	129
Private Sector Banks	11	15	10	11	36
Total	29	103	66	34	203
DCCB	1				35*
PACBs	185				185

Source: Annual Credit Plan 2000-2001, Vellore District.
* Branches of VDCCB

TRADE, COMMERCE AND EXPORT

Vellore is an important commercial centre in the state as well as world especially in leather goods. Vellore being the district headquarters town its lies midway in the Chennai, Bangalore National Highway and is well served by Road and Railway network. The various taluk headquarters and other towns are also well connected by good roads. The towns, such as Tiruppathur, Ambur, Vaniyambadi, Gudiyatham, Arcot and Ranipet are all prominent trade centres where many wholesale dealers and retailers dealing in cloth, grain and other articles.

Paddy, groundnut and tanned leather are the chief trades. Apart from agricultural products and other primary goods like foodgrain, leather goods, mangoes, vegetables, sandalwood, etc., manufactured goods like groundnut oil, sugarcane, jaggery, silk sarees, textile goods, beedi, woolen draggets, earthen jars, ceramic good's including sanitary-ware, leather articles, fertilizers and chemicals, bamboo and cane articles, korai mats etc., are marketed within as well as outside the district. Some of the raw materials for industries like the silk yarn, leaves for beedi rooling, etc., are imported into the district. Being an inland district, for exports foreign to countries, the district depend on Madras. Leather goods, beedies and matches are the most important commodity manufactured while leather goods and silk sarees are the most important commodity exported[8].

ADMINISTRATION

A. There are 3 administrative divisions functioning in the district :

(a) Ranipet (b) Vellore and (c) Tiruppathur

B. Taluk

The following 7 taluks are in Vellore district :

(a) Arakkonam, (b) Walajapet, (c) Gudiyatham, (d) Vellore, (e) Arcot, (f) Vaniyambadi, (g) Tiruppathur.

C. Community Development Blocks

There are 20 Community Development Blocks in Vellore districts. There are nine municipalities and 14 Town Panchayats.

D. Parliamentary constituencies

Vellore district contains 3 parliamentary constituencies, viz. :

(a) Arakkonam (b) Vellore and (c) Tiruppathur

The following Table helps in better understanding of the administrative position of the district.

Table 4.15
Particulars of District Administration

S.No.	*Particulars*	*Nos.*
1.	Total geographical area	5920 sq.kms
2.	Headquarters	Vellore
3.	No. of Taluks	8
4.	Revenue Divisions	3
5.	Community Divisional Blocks	20
6.	Municipal Towns	9
7.	No. of Town Panchayats	27
8.	No. of Village Panchayats	763
9.	No. of Revenue Villages	842

Source: Annual credit plan 2000-2001, Vellore District.

GENERAL CHARACTERISTICS OF SAMPLE VILLAGES PACBS

There are 185 primary agricultural co-operative societies functioning under the guidance and control of Vellore District Central Co-operative Bank. Two PACBs have selected for the study and have been identified after a through discussion with officials who have indicated the 'Best' and 'Bad'. Best is Thamaraipakkam PACB and Kalathure PACB is the 'bad', which is based on their performance. The official's opinion has been strengthened by primary and secondary data. However, officials have identified on the basis of some criteria; such as overdues, deposit mobilisation, issued loan amount, number of loan borrowed members etc.

The basic characteristic of these villages and PACBs are discussed in a separate section.

COVERAGE OF SAMPLE PACBS

The Thamaraipakkam PACB was registered on 27.6.1969 and it has been functioning from 6.9.1969. It has been given into "Best Performance" award among the district and the award was presented by Tamil Nadu State Co-operative Bank Ltd. during 1995-96. It is covers the following 6 villages to provide short term and medium-term loan.

The Kalathure PACB was registered on 3.2.1970 and functioning from 22.2.1970. The area covered under Kalathure PACB are 4 villages mentioned above.

Table 4.16 and 4.17 indicate that Thamaraipakkam PACB (2155.84 Hectare) is covering larger geographical area than the Kalathure PACB (1943.95 Hectares).

Table 4.16
Area Covered by Villages of Thamaraipakkam PACB

S.No.	*Villages*	*Area (Hectares)*
1.	Thamaraipakkam	736.41
2.	Perumanthangal	158.00
3.	Nangamangalam	128.39
4.	Vanakkambadi	335.33
5.	Mosur	591.73
6.	Palayam	205.98
	Total	2155.84

Source: Census of India 1991.

Table 4.17
Villages Covered Under Kalathure PACB

S.No.	*Villages*	*Area (Hectares)*
1.	Kalathure	958.37
2.	Avalur	411.68
3.	Perumpulipakkam	329.92
4.	Sankarambadi	243.98
	Total	1943.95

Source: Census of India 1991.

LOCATION

Thamaraipakkam PACB is located in Arcot taluk in Vellore district of Tamil Nadu. It is situated 8 kms south from Arani and 19 kms north from Arcot.

The Kalathure PACB is located in Arakkonam taluk of same district of Tamil Nadu. It is located 22 kms south from Arakkonam and 12 kms from Kaveripakkam. A state Highway runs for 7 kms along the full length of this area (east to west) from Madras to Vellore. The Madras-Bangalore broad gauge railway line cuts across the Arakkonam Taluk.

Katpadi and Arakkonam are the important Railway junctions in these study areas as well as the district. All the villages are electrified. All the villages covered under T PACB and K PACB have road facility.

Table 4.18A
Population in the Sample Villages of Thamaraipakkam PACB

S.No.	*Villages*	*1981*	*1991*	*Percentage Change*
1.	Thamaraipakkam	2207	2499	13.23
2.	Perumanthangal	553	555	0.36
3.	Nangamangalam	273	290	6.22
4.	Vanakkambadi	767	728	-5.08
5.	Mosur	1011	1042	3.06
6.	Palayam	767	773	0.78
	Total	5554	5887	5.99

Source: Census of India 1981, 1991.

Fig. 4.2: Vellore District Map

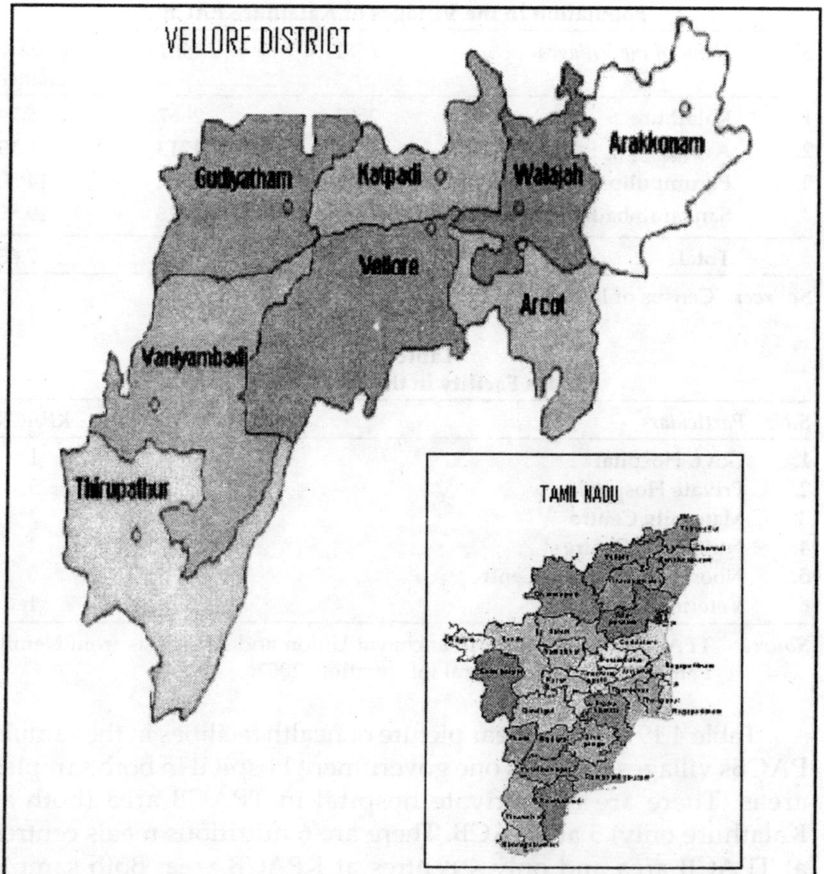

Table 4.18A and 4.18B reveal population growth of the sample PACBs villages in different census. It has been reduced by -5.08 per cent in Vanakkambadi village of TPACB and -1.15 per cent in Avalur village of Kalathure PACB during the year between 1981 to 1991. Except in these two villages (Vanakkambadi and Avalur) in all other villages there has been some marginal increased population. The census data reveal that villages of Thamaraipakkam PACB area have experienced less population growth than the Kalathure PACB villages. One can observe that a larger geographical area is covered by Thamaraipakkam PACB than Kalathure PACB.

Table 4.18B
Population in the Villages of Kalathure PACB

Sl.No.	*Name of the Villages*	*1981*	*1991*	*Percentage Change*
1.	Kalathure	2521	2667	5.79
2.	Avalur	1733	1713	-1.15
3.	Perumpulipakkam	1310	1504	14.80
4.	Sankarambadi	856	1026	19.85
	Total	6420	6910	7.63

Source: Census of India 1981, 1991.

Table 4.19
Health Facility in the Sample PACBs

S.No.	*Particulars*	*TPACB*	*KPACB*
1.	Govt. Hospital	1	1
2.	Private Hospital	2	5
3.	Maternity Centre	1	2
4.	Nutritious Centre	6	4
5.	Noon-meal Scheme Centre	5	6
6.	Veterinary Hospital	1	1

Source: TPACB is from Thimiri Panchayat Union and KPACB is from Nemili Panchayat Union, General information, 2002.

Table 4.19 shows a clear picture of health facilities in the sample PACBs villages. There is one government hospital in both samples areas. There are two Private hospital in TPACB area (both at Kalathure only) 5 at KPACB. There are 6 nutritious meals centres at TPACB area and only 4 centres at KPACB area. Both sample areas are having one Veterinary Hospital.

Table No.4.20 reports on the educational institutions in the sampled PACBs villages. All these villages are having Balwadies and only are Welfare school at Kalathure. There is no Higher secondary school in these villages. There is only are local General library at Thamaraipakkam.

Table 4.20
Educational Institutions in TPACB & KPACB

S.No.	Particulars	TPACB	KPACB
1.	Balwadies	7	5
2.	Primary & Middle School	6	6
3.	Welfare School	-	1
4.	High School	1	1
5.	Higher Secondary School	-	-
6.	Libraries	1	-

Source: TPACB from Thimiri Panchayat Union and KPACB from Nemili Panchayat Union, General information, 2002.

Table 4.21
Financial Institutions in the Sample PACBs

S.No.	Particulars	TPACB	KPACB
1.	PACB	1	1
2.	SBI	-	-
3.	Indian Bank	-	-
4.	Canara Bank	-	1
5.	Central Bank of India	1	-
6.	Co-operative Milk Society	2	3
7.	Post office	1	2
8.	Telephone Exchange office	1	-

Source: Information collected from V.A.O. Office records, 2002.

Table 4.21 indicates the position of financial institutions in sample villages. There is a branch of Canara Bank at Kalathure and Central Bank of India at Thamaraipakkam. 2 Co-operative Milk Society at Thamaraipakkam PACB villages (each at Thamaraipakkam and Mosur) and 3 at Kalathure PACB villages (are at Kalathure, Avalur and Perumpulipakkam). Thamaraipakkam has a Telephone Exchange office.

Table 4.22
Sources of irrigation in the sample PACBs

S.No.	*Particulars*	*TPACB*	*KPACB*
1.	Well with pumpsets	675	450
2.	Well with oil engine	-	8
3.	Tank	8	2
4.	Lake	6	4
5.	Canals & Others	-	1

Source: Information collected from V.A.O. Office records, 2002.

Table 4.22 explains the sources of irrigation facilities in the sample PACBs villages. Agricultural operations in these area is mainly dependent on the seasonal rainfall. Well with pumpsets is the major sources of irrigation in the sample villages. There is a canal sources in the area of Kalathure only.

CROPPING PATTERN

Both PACB villages cultivate for three agricultural seasons viz. "*Sornavari*" (May to September), "*Samba*" (August to February) and "*Navarai*" (December to May) and mostly are suited for paddy cultivation.

Table 4.23
Area Under Cultivation of Selected Crops in Thamaraipakkam PACB Villages During 2001-2002

(in Hectare)

Name of the villages	*Paddy*	*Cholam*	*Cumbu*	*Ragi*	*Sugar-cane*	*Ground-nut*	*Banana*	*Total*
Thamarai-pakkam	210 (37.84)	15 (2.70)	-	10 (1.80)	250 (45.05)	50 (9.00)	20 (3.61)	555
Peruman-thangal	20 (16.00)	-	-	-	60 (48.00)	45 (36.00)	-	125
Nangaman-galam	15 (16.85)	-	-	4 (4.49)	60 (67.42)	10 (11.24)	-	89
Vanakkampadi	80 (31.37)	10 (3.92)	20 (7.84)	-	110 (43.15)	15 (5.88)	20 (7.84)	255
Mosur	100 (27.40)	25 (6.85)	10 (2.74)	20 (5.48)	90 (24.66)	50 (13.69)	70 (19.18)	365
Palayam	90 (50.00)	-	-	-	7 (38.90)	10 (5.55)	10 (5.55)	117
Total	515 (34.20)	50 (3.32)	30 (1.99)	34 (2.24)	577 (38.31)	180 (11.95)	120 (7.97)	1506 (100.00)

Source: Information collected from V.A.O. Office records, 2002.

Table 4.23 and 4.24 give the information about the area under cultivation and selected crops cultivated in the sample PACBs villages in the year 2001 2002. It is shown in the table that paddy and sugarcane are the most important crops in terms of area under cultivation. Groundnut is another crop, which is cultivated in considerable area. Banana is cultivate more (in terms of area) in Thamaraipakkam PACB villages than Kalathure PACB villages.

Table 4.24
Area Under Cultivation of Selected Crops in Kalathure PACB Villages During 2001-2002

(in Hectare)

Name of the villages	*Paddy*	*Cholam*	*Cumbu*	*Ragi*	*Sugar-cane*	*Ground-nut*	*Banana*	*Total*
Kalathure	202	-	-	4	25	50	23	304
	(66.45)			(1.31)	(8.22)	(16.45)	(7.84)	
Avalur	170	-	-	3	50	10	-	233
	(72.96)			(1.29)	(21.46)	(4.29)		
Perumpuli-pakkam	110	5	10	-	20	10	-	155
	(70.97)	(3.23)	(6.45)		(12.90)	(6.45)		
Sankarambadi	130	-	-	20	-	30	10	190
	(68.42)			(10.53)		(15.79)	(5.26)	
Total	612	5	10	27	95	100	33	882
	(69.39)	(0.57)	(1.13)	(3.06)	(10.77)	(11.34)	(3.74)	(100.00)

Source: Information collected from V.A.O. Office records, 2002.

One can observe that paddy is more in Kalathure PACB village (in terms of area) than in Thamaraipakkam villages. One can see that Palar river is passing nearby Sankarambadi village which is a part of Kalathure PACB.

Thus the entire analysis reveals that the villages covered under both PACBs are more or less in the same position and the district has shown considerable progress over the years in every respect.

REFERENCES

1. Government of Madras, Manual of Administration of Madras Presidency, Madras, 1895, p. 53.
2. Government of India: Census Hand Books of North Arcot, 1991, p.1.
3. Op.cit. Government of Madras. p. 53.

4. Govindarajan, P., 'Economics of Poverty, Rural Poverty an Institutional Approach, with reference to North Arcot District University of Madras, Madras, 1984, p.178.

5. Ramasamy, C., Hazell, B.R., Peter and Aiyasamy, P.K., North Arcot and Green Revolution, Ch.2, In Peter B.R.Hazell and Ramasamy, C., (eds) *The Green Revolution Reconsidered, The Impact of High Yielding Rice varieties in South India,* The International Food Policy Research Institute, Oxford University Press, Oxford, 1991, p.15.

6. Potential linked credit plan, Vellore District 2001-2002, NABARD, p. 39.

7. Government of Tamil Nadu, Website: www.tn.gov.in.

8. Sivaprakasam. R. "Housing conditions of scheduled caste in Vellore District (Tamil Nadu) : A study, University of Madras, Chennai, 2000, p. 86.

9. Annual Credit Plan, 2000-2001, Vellore District. p.1.

10. District Collectorate, Profile of Vellore District, Vellore, 9.

11. Census of India 1991, Tamil Nadu Series 20, Part III, A and B, p. 107-123.

Chapter - 5

WORKING OF CO-OPERATIVE CREDIT MOVEMENT

This chapter presents an assessment of the supply of the credit to the farmers based on the secondary data collected from State Co-operative Bank, Vellore District Central Co-operative Bank, Thamaraipakkam PACB and Kalathure PACB.

The secondary data pertaining to year-wise total membership and actual loan user, year-wise loan distribution of short term, medium-term and others, recovery performance, deposit mobilisation and other characteristics relating to the co-operative banks are summarised and analysed below:

Table 5.1 shows the flow of institutional credit to agricultural sector. Co-operatives have played a significant role in the flow of credit to the agricultural sector, both short-term as well as medium-term loan. Apart from substantial expansion in the flow of credit from commercial banks and Regional Rural Banks (RRBs), the co-operative credit institutions continue to be an important institutional agency for providing credit support and services for agricultural and rural development. Agency-wise disbursement of agricultural credit from 1989-1990 to 2001-2002 is given in Table 5.1. The flow of institutional credit for agriculture and allied activities has increased from Rs.10628.2 crore in 1989-1990 to Rs. 66771 crore in 2001-2002.

Today, co-operatives form the major component of the rural credit system in India. Of the total credit (including short term, medium term and long-term) disbursed during the year 1989-90, the share of co-operatives was 50.87 percent, i.e., about half of the total credit disbursed. The share of the commercial banks and Regional Rural Banks (RRBs), the other two institutional players

Table 5.1
Flow of Institutional Credit to Agriculture Sector in India

(Rupees in crore)

Sl. No.	Name of the Institutions	1989-90	1990-91	1991-92	1992-93	1993-94	1994-95	1995-96	1996-97	1997-98	1998-99	1999-2000	2000-2001**	2001-2002***
1.	Co-operative Banks	5406.8	4819.1	5796.8	6483.7	8484.0	9875.5	12483.0	11944	14085	15957	18363	21909	27080
		(50.87)	(47.30)	(50.24)	(51.75)	(56.51)	(52.60)	(52.69)	(45.78)	(44.08)	(43.29)	(39.63)	(40.96)	(40.56)
2.	Commercial Banks and Regional Rural Banks	5221.4	5368.6	5740.7	6046.1	6529	8897.6	11208.9	14467	17871	20903	27905	31595	39691
		(49.13)	(52.70)	(49.76)	(48.25)	(43.49)	(47.40)	(47.31)	(54.78)	(55.92)	(56.71)	(60.31)	(59.05)	(59.44)
	Total	10628.2	10187.7	11537.5	12529.8	15013	18773.1	23691.9	26411	31956	36860	46268	53504	66771
	Per cent increase	12.30	-4.14	13.25	8.60	20	25	26	11.48	21	15	26	20	25

Source: 1. Hand Book of statistics on Indian Economy RBI - 1999, p.68.
2. Economic Survey, 2001-2002.

Note: Figures in brackets are percentage to total
** Estimated
*** Projections

in the rural credit, together was 49.13 percent of the total disbursement of agricultural credit. Though the share of co-operatives (i.e. 40.56 per cent) in agricultural credit had shown a declining trend till 2001-2002. It has shown strong signs or revival in the year 1993-1994.

A meaningful evaluation of the performance of TNSC bank in Tamil Nadu, some important indicators are identified. An examination of these indicators as given in Table 5.2 shows that the bank has made all round progress. The paid-up capital was reached by Rs.5.58 crores in 1989-90 to Rs.27.28 crores in 1999-2000. Similarly, deposits of the bank increased from Rs.387.24 crores in 1989-90 to Rs.2676.11 crores in 1999-2000. The fresh loans advanced by TNSC banks reached a level of (ST only) Rs.62020.38 lakhs in 1998-99 from a mere Rs.49121.00 lakhs in 1997-98 regarding a growth of 1.3 times.

Table No. 5.2a shows that the particulars of Central Co-operative Bank in Tamil Nadu from 1993-94 to 1999-2000. The total number of District Central Co-operative Bank (DCCB) increased from 20 in 1993-94 to 23 in 1999-2000. The Deposits with DCCB went up from Rs. 1680.22 crores in 1993-94 to Rs. 6022.06 crores in 1999-2000. The Short-term agricultural loans issued through the DCCBs increased from Rs. 138.93 crores in 1993-94 to Rs. 879.21 crores during the period 1999-2000. Similarly, Medium-term agricultural loans were reached by Rs. 100.57 crores during the period from 1993-94 to 1999-2000.

There has been a significant progress in the lending operations of the District Central Co-operative Banks in Tamil Nadu.

THE CO-OPERATIVE MOVEMENT IN TAMIL NADU

The co-operative movement, which has already made good progress in various fields of economic activities in this state is now passing through a period of accelerated expansion.[1] The state has built up a well-defined and strong co-operative credit structure.

Tamil Nadu (formerly known as Madras State) takes pride of place in co-operation in our country. It was in then Madras that the foundation of co-operative was laid in 1903. The true pioneers were Shree Srinivasa Sastri and his colleagues who started in 1903 the Triplicane Urban Co-operative Credit Society Ltd., which has not only survived but has become the centre of inspiration for consumer's co-operation in the whole of South-East Asia. The first

Table 5.2
Key Indicators of Progress of the State Co-operative Bank - At a Glance

(Rupees in crores)

Sl. No.	*Indicators*	*1989-1990*	*1990-1991*	*1991-1992*	*1992-1993*	*1993-1994*	*1994-1995*	*1995-1996*	*1996-1997*	*1997-1998*	*1998-1999*	*1999-2000*
1.	Number of TNSC Banks	1	1	1	1	1	1	1	1	1	1	1
2.	Paid-up capital	5.58	5.58	5.83	5.83	6.33	11.53	17.76	18.74	18.74	21.90	27.28
3.	Total owned funds	55.48	67.44	68.30	75.03	100.89	113.79	140.93	170.01	181.36	193.80	204.93
4.	Deposits	387.24	415.29	451.58	550.50	661.21	760.29	938.53	1213.93	1634.58	2239.60	2676.11
5.	Borrowings	311.36	476.32	469.34	469.99	354.79	787.62	694.00	412.31	390.45	647.85	493.83
6.	Working capital	789.33	1026.56	1054.03	1160.08	1174.02	1726.33	1858.09	1881.66	2306.07	3186.72	3502.74
7.	Advances out standing	599.72	794.51	800.75	848.37	914.53	1371.13	1318.46	1379.05	1393.40	1261.63	1199.66
8.	Net profit	9.63	10.10	12.63	15.39	8.10	22.46	35.11	18.43	15.07	2.51	9.74

Source: 94th Annual Report and Statement of Accounts. The Tamil Nadu State Apex Co-operative Bank Ltd. 1999-2000, Chennai - 1.

Table 5.2a
Particulars of Central Co-operative Banks in Tamil Nadu

(Rupees in crores)

Sl. No.	*Particulars*	*1993-94*	*1994-95*	*1995-96*	*1996-97*	*1997-98*	*1998-99*	*1999-00*
1.	Number of Central Co-operative Banks	20	22	22	23	23	23	23
2.	Number of Branches	590	642	657	663	686	708	716
3.	Deposits Held	1680.22	1923.85	2360.97	2965.03	3736.91	4884.17	6022.06
4.	Advances Outstanding	1812.96	2856.04	3147.69	3718.39	4126	4463.43	5008.81
5.	Short-term Agricultural Loans Issued	138.93	561.73	557.86	590.69	697.99	732.12	879.21
6.	Medium-term Agricultural Loans Issued	11.08	56.04	67.05	63.42	96.76	117.13	111.65

Source: Tamil Nadu—An Economic Appraisal 2001-2002.

co-operative land mortgage bank was started in Punjab but the foundations of this branch were laid in Tamil Nadu of all other parts it is in Madras that co-operative land mortgage banking has taken root and flourished. In harnessing co-operation even for other economic activities, Tamil Nadu has not lagged behind other states. Sri.P. Rajagopalachari was the first Registrar of Co-operative Societies in Tamil Nadu and he was appointed in July 1904 and the first society to be registered was the Tirur Rural Society in Thiruvallur district.[2]

Tamil Nadu in course of time had seen several agricultural co-operative societies come into existence. The central co-operative banks in Tamil Nadu form the backbone of the financing system of the co-operative movement. The Madras Central Urban Bank was the first bank in Tamil Nadu registered on October 19, 1905 under Act of 1904, with the sole object of finding money to finance societies in all parts of the state. In accordance with the recommendations of the Maclagan Committee on co-operation 1915, the Madras Central Urban bank was converted into an Apex State Co-operative bank in April 1917. The constitution of the State Co-operative Bank was perfected in 1919 by making it federation of central banks with individuals also as shareholders and by transferring the shares held by the primary societies to central bank. Thereafter the state co-operative bank ceased to finance primary societies in the districts directly.[3] The central co-operative banks are the federations of primary agricultural co-operative banks and co-operatives of other types in their respective areas.

During the 1920 for provision of long-term agricultural credit, land mortgage co-operative banks were established first in the than Bombay province followed by Tamil Nadu. These were primary banks, the need for a central apex institution to affiliate these was felt by the government of Tamil Nadu. In 1928, the Madras Central Co-operative Land Mortgage Bank was started in Tamil Nadu. It was a matter of numerical expansion during 1930s and 1940s for the PACBs, urban banks, land mortgage banks and central banks. It was only after the All-India Rural Credit Survey Committee Report of 1956, that real expansion in all spheres of co-operative credit took place in Tamil Nadu. The period, 1960 and 1970 witnessed an alround growth; during the 1980s deposits of central banks and urban banks increased. But overdues became very heavy in short-term and long-term credit sectors which trend continued in the beginning of 1990 also.[4] As on March 31 2001, there were 23 central co-operative banks in Tamil Nadu.

The constitution has undergone various modifications from time to time to meet the growing demands of the movements.

The Tamil Nadu State Apex Co-operative bank, has bagged the first prize for the first time for its best performance instituted by NABARD for the year 2000-2001 among all the State Co-operative banks (SCBs) in the country. The apex bank was earlier awarded second prize for the year 1995-1996 and 1998-1999 for its best performance. At the state level, the Tiruchirappalli district central co operative bank and the Salem district central co-operative banks have been awarded the first and second prizes respectively for the year 2000-2001 among the district central co-operative banks in the state.[5]

GENERAL PROGRESS OF VELLORE DISTRICT CENTRAL CO OPERATIVE BANK

The Vellore district central co-operative bank was registered on 06.05.1917. The member societies in the year 1917-1918 were individual by 48 and societies by 295. During the year 1967-68 individuals were 126 and societies by 1446. But during 2001-2002, 629 member societies were in VDCCB, out of this 185 societies are PACBs.

The share capital in the year 1917-1918 was only Rs.75,712 and it has increased in the year 2001-2002 to Rs.10.76 crores. Deposits of the Vellore district central co-operative bank also in the year 1917-1918 was Rs.3,10,938 and in the year 2001-2002 was Rs. 275.73 crores. The loan outstanding was only Rs.3,65,524 in the year 1917-1918 and it has increased to Rs.321.00 crores in the year 2001-2002.

In the year 1917-1918, profit earned was Rs.6337 and no dividend was given. During the period 1957-1958 profit were Rs.85,061 and dividend was given at 4.50 percent. Further, in the year 1994-1995, there been neither profit, nor loss. In the year 2000-2001 in the society earned a loss of Rs. 369.69 crores. But profit earned is Rs.4.00 lakhs during the year 2001-2002.[6]

Table 5.3 gives the information regarding the details of ground level credit disbursement under agriculture and allied activities by co-operative banks and commercial banks in Vellore district. The co-operative credit societies occupy important place in disbursement of agricultural credit among the institutional agencies. Crop loan is for a period of not exceeding 16 months and it meets only cultivation purposes. Generally, it is repaid after cultivation of crops. The role of co-operative banks as well as commercial banks for the distribution of agricultural credit has steadily grown up in the district. On the other hand, institutional

Table 5.3
Details of Ground Level Credit Disbursed by Institutional Agencies in Vellore District

(Rupees in crores)

Sl.No.	*Year*	*Co-operative Banks*	*Commercial Banks*	*Total*
1.	1988-1989	52.38	18.40	70.78
		(74.00)	(26.00)	(100.00)
2.	1989-1990	NA	NA	NA
3.	1990-1991	NA	NA	NA
4.	1991-1992	17.14	26.54	43.68
		(39.24)	(60.76)	(100.00)
5.	1992-1993	18.36	43.59	61.95
		(29.64)	(70.36)	(100.00)
6.	1993-1994	16.40	49.69	66.09
		(24.81)	(75.19)	(100.00)
7.	1994-1995	18.06	55.36	73.42
		(24.60)	(75.40)	(100.00)
8.	1995-1996	22.85	66.99	89.84
		(25.43)	(74.57)	(100.00)
9.	1996-1997	35.59	125.69	161.28
		(22.07)	(77.93)	(100.00)
10.	1997-1998	56.71	126.70	183.41
		(30.92)	(69.08)	(100.00)
11.	1998-1999	44.45	120.38	164.83
		(26.97)	(73.03)	(100.00)
12.	1999-2000	70.35	132.81	203.16
		(34.63)	(65.37)	(100.00)
13.	2000-2001	62.13	126.21	188.34
		(32.99)	(67.01)	(100.00)
14.	2001-2002	84.52	115.07	199.59
		(42.35)	(57.65)	(100.00)

Source: Various Annual Credit Plans, Vellore District
Note: Figures in brackets are percentage to total.

agencies are fully involved in their activities on production oriented, namely minor irrigation, land development, farm mechanisation, plantation and horticulture, development of dairy, poultry, sheep/goat/piggery, forestry and waste land development, bio-gas, sericulture and others. During 1988-89, the share of crop loans and allied activities through by commercial banks was 26.00 per cent and it rose to 57.65 per cent by 2001-2002. Co-operatives account for 74.00 percent in 1988-89 and declined to 42.35 per cent for 2001-2002.

Table 5.4
Particulars of Total Membership and Loan Borrowers in VDCCB

Sl. No.	*Years*	*Total Members*	*Actual Loan Users*	*Balance*
	1	2	3	4
1.	1989-1990	NA	NA	NA
2.	1990-1991	NA	NA	NA
3.	1991-1992	827240	320545	506695
		(100.00)	(38.75)	(61.25)
4.	1992-1993	794482	330334	464148
		(100.00)	(41.58)	(58.42)
5.	1993-1994	806860	337454	469406
		(100.00)	(41.82)	(58.18)
6.	1994-1995*	423346	126211	297135
		(100.00)	(29.81)	(70.19)
7.	1995-1996	444799	148737	296062
		(100.00)	(33.44)	(66.56)
8.	1996-1997	463746	165674	298072
		(100.00)	(35.73)	(64.27)
9.	1997-1998	483269	176972	306297
		(100.00)	(36.62)	(63.38)
10.	1998-1999	505774	197251	308523
		(100.00)	(39.00)	(61.00)
11.	1999-2000	531184	198054	333130
		(100.00)	(37.29)	(62.71)
12.	2000-2001	521926	211279	310647
		(100.00)	(40.48)	(59.52)
13.	2001-2002	532995	213071	319924
		(100.00)	(39.98)	(60.02)

Source: Annual Report of VDCCB Vellore
Note: Figures in brackets indicate percentage to Total

*During 1994-95 Vellore District Central Co-operative Bank was divided into the Tiruvannamalai District Central Co-operative Bank and Vellore District Central Co-operative Bank.

Total Membership and Actual loan user

Multiple Regression on Total loan users

Multiple regression model has been fitted to the data considering number of actual loan users as dependent variable and total members in vellore central co-operative Bank as independent variable.

R Square	=	0.98180
F Value	=	323.67**

Variables in the Equation

Variable	B	SE B	Beta	T Value	P Value
Total Members	0.486	0.027	0.990	17.991	0.000**
(Constant)	63445.039	16692.153		3.801	0.009**

Statistics for the model has a significance level of 0.0000, which tells that at least one of the correlations between each of the independent variables and the dependent variable is not equal to zero in the population. The model is best fit for the data.

Multiple regression co-efficient of determination (R square) value is .982, which showing that 98 percent of variations in the dependent variable can be explained by independent variable.

The co-efficient of total members is 0.4865, which shows that for every one unit increase in number of members in the PACB, there is a 0.49 unit increase in the number of loan users.

Table 5.4 gives information regarding total members and actual loan users during 1991-1992 to 2001-2002. There are 827240 and it has decreased to 806860 in 1993-1994. Actual loan users in the year 1991-1992 is 320545 (i.e 38.75 percent of the total members) and it has increase to 337454 (41.82 percent) by the year 1993-1994. In the year 1994-1995 Vellore Central Co-operative Bank has been bifurcated into vellore central co-operative bank and Tiruvannamalai central co-operative bank. Data given in the table, from 1994-1995 is only for VDCCB. In the year 1994-1995 there were 423346 members in VDCCB and it has increased by 109649 members. In the same time actual loan users increased from 126211 members in the year 1994-1995 to 213071 members in the year 2001-2002.

Table 5.5 gives a clear idea about particular of loans issued by all PACBs in vellore District. It consists of short-term (ST), medium-term (MT), medium-term conversion (MTC) and others. Others includes like, jewells loan, consumers loan, housing loan and loan on deposit (LOD). In the year 1990-1991 the loan amount issued by all PACB in vellore district was Rs.2802.17 lakhs and it has increased to Rs.19717 lakhs by the year 2001-2002. In the year 1990-1991 short-term loan accounts for 53.74 percentage of the total loan issued. Percentage of the short-term loan in the total loan amount has been reduced to 47.57 percent by the year 2001-2002.

Table 5.5
Particulars of Loan Issued During 1989-1990 to 2001-2002 to all PACBs in Vellore District by VDCCB

(Rupees in Lakhs)

Sl. No.	*Years* 1	*ST* 2	*MT* 3	*MTC* 4	*Others* 5	*Total* 6
1.	1989-1990	NA	NA	NA	NA	NA
2.	1990-1991	1505.80	180.60	53.03	1062.74	2802.17
		(53.74)	(6.44)	(1.89)	(37.93)	(100.00)
3.	1991-1992	1663.36	198.91	12.67	817.95	2692.89
		(61.77)	(7.39)	(0.47)	(30.37)	(100.00)
4.	1992-1993	1848.53	251.90	16.12	1090.04	3206.59
		(57.65)	(7.86)	(0.50)	(33.99)	(100.00)
5.	1993-1994	2191.36	311.97	37.06	1511.96	4052.35
		(54.08)	(7.70)	(0.91)	(37.31)	(100.00)
6.	1994-1995	3351.82	378.57	28.10	2216.55	5975.04
		(56.10)	(6.33)	(0.47)	(37.10)	(100.00)
7.	1995-1996	3232.78	354.85	41.65	2641.00	6270.28
		(51.56)	(5.66)	(0.66)	(42.12)	(100.00)
8.	1996-1997	4504.53	577.56	60.87	3649.20	8792.16
		(51.23)	(6.57)	(0.69)	(41.51)	(100.00)
9.	1997-1998	5074.75	499.79	84.67	4726.37	10385.58
		(48.86)	(4.81)	(0.82)	(45.51)	(100.00)
10.	1998-1999	4502.28	625.53	163.07	5563.69	10854.57
		(41.48)	(5.76)	(1.50)	(51.26)	(100.00)
11.	1999-2000	5009.45	405.24	182.07	7267.79	12864.55
		(38.94)	(3.15)	(1.42)	(56.49)	(100.00)
12.	2000-2001	7483.4	1422.95	364.31	8659	17929.66
		(41.74)	(7.94)	(2.03)	(48.29)	(100.00)
13.	2001-2002	9378.54	728.17	178.22	9432.14	19717.07
		(47.57)	(3.69)	(0.90)	(47.84)	(100.00)

Source: Annual Report of VDCCB Vellore.
Note: Figures in brackets are percentage to total.

Accumulation of overdues constitutes a major syndrome affecting the sound functioning of co-operative banks. The real problem prevails at the primary level, where the co-operative banks have to recover the dues from the beneficiaries. Table No.5.6 shows that overdues percentage was 52.03 percent in 1990-91 which decreased to 34.00 percent in 2001-2002. It was at maximum 54.84 percent in the year 1995-96. It can be seen from the table that the overdues in Vellore Central Co-operative Bank is continuously on a downward trend since the bank has taken strict measures for reducing the percentage of overdues.

Table 5.6
Recovery Performance of VDCCB

(Rs. in lakhs)

Sl. No.	*Years*	*Demand*	*Collection*	*Overdues*	*Percentage of overdue to demand*
	1	2	3	4	5
1.	1989-1990	NA	NA	NA	-
2.	1990-1991	3533.25 (100.00)	1694.71 (47.96)	1838.54 (52.04)	52.03
3.	1991-1992	4131.51 (100.00)	1865.40 (45.15)	2266.11 (54.85)	54.84
4.	1992-1993	4380.59 (100.00)	2173.98 (49.63)	2206.61 (50.37)	50.37
5.	1993-1994	4912.28 (100.00)	2588.44 (52.69)	2323.84 (47.31)	47.30
6.	1994-1995	6031.69 (100.00)	3572.22 (59.22)	2459.47 (40.78)	40.77
7.	1995-1996	6448.67 (100.00)	3793.02 (58.82)	2655.65 (41.18)	41.18
8.	1996-1997	7544.33 (100.00)	4638.55 (61.48)	2905.78 (38.52)	38.51
9.	1997-1998	8955.73 (100.00)	5387.54 (60.16)	3568.19 (39.84)	39.84
10.	1998-1999	10031.78 (100.00)	6073.96 (60.55)	3957.82 (39.45)	39.45
11.	1999-2000	11280.93 (100.00)	7215.36 (63.96)	4065.57 (36.04)	36.03
12.	2000-2001	12052.45 (100.00)	7767.13 (64.44)	4285.32 (35.56)	35.55
13.	2001-2002	11403.90 (100.00)	7526.43 (66.00)	3877.47 (34.00)	34.00

Source: Various Annual Reports of VDCCB, Vellore.
Note: Figures in brackets are percentage to total.

Table 5.7
Yearwise Particulars of Deposit Position by Member PACBs

(Rupees in lakhs)

Sl. No.	*Years*	*Amount of Deposit*	*Indices*
	1	2	3
1.	1989-1990	NA	-
2.	1990-1991	529.84	100.00
3.	1991-1992	757.67	143.21
4.	1992-1993	863.49	163.21
5.	1993-1994	1080.96	204.32
6.	1994-1995	1628.76	307.87
7.	1995-1996	2608.71	493.10
8.	1996-1997	4143.41	783.19
9.	1997-1998	5352.47	1011.73
10.	1998-1999	7371.15	1393.80
11.	1999-2000	10022.37	1891.58
12.	2000-2001	11383.29	2148.43
13.	2001-2002	12216.07	2305.61

Source: Various Annual Report of VDCCB, Vellore.

Deposits are one of the important sources of working capital of a co-operative banks. The mobilisation of deposits is essential for the development of business and for meeting the financial requirements of primary agricultural co-operative banks. The central co-operative banks mobilise various kinds of deposits from the members and non-members in the district. Increased deposits will enable the bank to reduce its dependence on external borrowings. It is the size of the deposits that largely decides the lending activities of a bank. Thus, the importance of deposit mobilisation for a bank needs much emphasis.

It can be concluded from the data given in Table No.5.7 that amount of deposit has been consistently increasing. It was Rs.529.84 lakhs in the year 1990-1991 and has increasedby Rs.11686.23 lakhs to reach Rs.12216.07 lakhs by the year 2001-2002.

Considering 1990-1991 as base years one can observe that the deposits have increased by 22 times. It is clear that deposits of all PACB membership at VDCCB, has grown up considerably over the 12 years period.

Table 5.8
Correlation co-efficient Relationship Between Deposit, Total Membership and loan issued

Sl.No.	*Years*	*Deposit*	*Total membership*	*Loan issued*
	1	2	3	4
1.	Deposit	1.000	-.6069 P = .111	9630 .P = .000*
2.	Total membership	-.6069 P = .111	1.0000	-.7598** P = .029
3.	Loan issued	.9630 P = .000*	-.7598 P = .029*	1.0000

Source: Secondary data.
Note: *Significant at 1% level.
** Significant at 5% level.

Correlation co-efficient has been given in Table No.5.8. The relationship between deposits and loans issued is significant at 1% level and the correlation co-efficient is as high as .96. The relationship between total members and loans issued is significant at 5% level. It has a co-efficient value of -.7598, which denotes a high negative relationship.

Table 5.9
Year wise Particulars of Overdues in VDCCB for all PACBs
(Rupees in lakhs)

Sl.No.	*Years*	*Amount of overdues*	*Indices*
1.	1989-1990	NA	NA
2.	1990-1991	1838.54	100
3.	1991-1992	2266.11	123.25
4.	1992-1993	2206.61	120.10
5.	1993-1994	2323.84	126.39
6.	1994-1995*	2459.47	133.77
7.	1995-1996	2655.65	144.44
8.	1996-1997	2905.78	158.04
9.	1997-1998	3568.19	194.07
10.	1998-1999	3957.82	215.26
11.	1999-2000	4065.57	221.13
12.	2000-2001	4285.32	233.08
13.	2001-2002	3877.47	210.90

Source: Various Annual Report of VDCCB Vellore.

Considering 1990-1991 as base year, one can observe that the over dues has increased by 1½ times. That the amount of overdues has been consistently increasing through the PACB. It was Rs.1838.54 lakhs in the year 1990-1991 and has increased by Rs. 2038.93 lakhs to reach Rs. 3877.47 lakhs by the year 2001-2002.

General Progress of Thamaraipakkam PACB and Kalathure PACB

Since the size of land holdings has an important role in determining the size of loan, an attempt has been made to prepare a table on the land wise distribution of members. The members of a society are the pillars on which the co-operative credit movement rests.

Table 5.10 shows the growth of membership in Thamaraipakkam PACB. A perusal of the above table clearly brings out the fact that farmers who are holding land between 2 to 4 Hectares constitute the declining 30.81 percent in the year 1989-1990 to only 17.53 percent in the year 2001-2002. But farmers, who belong to the below 1 hectare group steadily increased from 15.08 percent in 1989-1990 to 27.49 percent in 2001-2002. This confirms that higher percentage coverage of borrowing members were from the land holders belonging to below 1 hectare. This is a measure of the good performance of Thamaraipakkam PACB which helped the poor and needy farmers.

Table 5.10a clearly shows the growth of membership in Kalathure PACB in 1990-1991, 1991-1992 and 1994-1995 data is not available for the study period. Actually the proportion of members has declined from 32.90 percent in 1989-1990 to 29.87 percent in the terminal year 2001-2002.

The proportion of members in the upper size groups more or less remain in the same group over the years with minor changes in both the societies, which reflects that landless Agricultural labourers are registered much higher in Kalathure PACB (15.89 percent) than Thamaraipakkam PACB (4.19 per cent).

Thus, from the above analysis, it can clearly be inferred that the contribution of membership by the landholders below 4 hectare is higher than that of the land holders having 4 hectare and above. This also shows that a greater number of small land holders (i.e, below 4 hectare) are benefitted through both the societies.

Table 5.10
Status of Membership in Thamaraipakkam PACB

Land Size	*1989-1990*	*1990-1991*	*1991-1992*	*1992-1993*	*1993-1994*	*1994-1995*	*1995-1996*	*1996-1997*	*1997-1998*	*1998-1999*	*1999-2000*	*2000-2001*	*2001-2002*
2	3	4	5	6	7	8	9	10	11	12	13	14	15
Below 1	229	239	108	148	265	269	275	375	399	399	469	429	439
Hectare	(15.08)	(15.37)	(13.37)	(14.12)	(23.75)	(23.75)	(21.42)	(25.61)	(27.01)	(26.09)	(27.79)	(27.22)	(27.49)
1-2	328	338	229	279	339	339	349	389	392	445	445	435	438
	(21.59)	(21.74)	(28.34)	(26.62)	(30.38)	(30.38)	(27.18)	(26.57)	(26.54)	(29.08)	(26.36)	(27.60)	(27.43)
2-4	468	468	183	223	244	244	249	249	274	274	274	274	280
	(30.81)	(30.10)	(22.65)	(21.28)	(21.86)	(21.86)	(19.39)	(17.01)	(18.55)	(17.91)	(16.23)	(17.39)	(17.53)
4-8	114	124	47	87	42	42	45	45	75	75	75	75	75
	(7.51)	(7.97)	(5.82)	(8.30)	(3.76)	(3.76)	(3.50)	(3.07)	(5.08)	(4.90)	(4.44)	(4.76)	(4.69)
8 Hectare &	57	58	57	57	47	47	47	47	14	14	14	14	14
above	(3.75)	(3.73)	(7.05)	(5.44)	(4.21)	(4.21)	(3.66)	(3.21)	(0.95)	(0.91)	(0.83)	(0.89)	(0.88)
Agri. labour	285	285	63	123	63	63	65	125	240	240	290	260	262
	(18.76)	(18.33)	(7.79)	(11.74)	(5.65)	(5.65)	(5.06)	(8.54)	(16.25)	(15.69)	(17.18)	(16.50)	(16.41)
Rural Artisan	38	38	13	23	13	13	20	120	24	24	24	22	22
	(2.50)	(2.44)	(1.61)	(2.19)	(1.16)	(1.16)	(1.56)	(1.37)	(1.63)	(1.57)	(1.42)	(1.39)	(1.38)
LAL	NA	5	108	108	103	103	234	214	59	59	97	67	67
		(0.32)	(13.37)	(10.31)	(9.23)	(9.23)	(18.23)	(14.62)	(3.99)	(3.86)	(5.75)	(4.25)	(4.19)
Total mem-	1519	1555	803	1048	1116	1223	1284	1464	1477	1530	1688	1576	1597
bership	(100.00)	(100.00)	(100.00)	(100.00)	(100.00)	(100.00)	(100.00)	(100.00)	(100.00)	(100.00)	(100.00)	(100.00)	(100.00)

Source: Various annual reports of TPACB.
Note : Figures in brackets are Percentage to total.

Table 5.10a
Status of Membership in Kalathure PACB

Land Size	*1989-1990*	*1990-1991*	*1991-1992*	*1992-1993*	*1993-1994*	*1994-1995*	*1995-1996*	*1996-1997*	*1997-1998*	*1998-1999*	*1999-2000.*	*2000-2001*	*2001-2002*
2	3	4	5	6	7	8	9	10	11	12	13	14	15
Below 1	563	NA	NA	595	610	NA	625	640	640	650	655	655	705
Hectare	(32.90)			(31.45)	(31.52)		(30.31)	(29.89)	(29.89)	(29.80)	(29.80)	(29.77)	(29.87)
1-2	346	NA	NA	367	377	NA	389	399	399	413	417	419	449
	(20.22)			(19.40)	(19.48)		(18.87)	(18.64)	(18.64)	(18.93)	(18.97)	(19.05)	(19.02)
2-4	189	NA	NA	186	194	NA	198	208	208	215	215	215	225
	(11.05)			(9.83)	(10.03)		(9.60)	(9.71)	(9.71)	(9.86)	(9.78)	(9.77)	(9.53)
4-8	120	NA	NA	145	155	NA	159	164	164	167	167	167	167
	(7.01)			(7.66)	(8.81)		(7.71)	(7.66)	(7.66)	(7.66)	(7.60)	(7.59)	(7.08)
8 Hectare &	28	NA	NA	40	40	NA	40	40	40	32	32	32	32
above	(1.64)			(2.11)	(2.07)		(1.94)	(1.87)	(1.87)	(1.47)	(1.45)	(1.46)	(1.36)
Agri. labour	181	NA	NA	214	214	NA	257	262	262	270	273	273	293
	(10.58)			(11.31)	(11.06)		(12.46)	(12.24)	(12.24)	(12.38)	(12.42)	(12.41)	(12.42)
Rural Artisan	72	NA	NA	100	100	NA	109	114	114	114	114	114	114
	(4.21)			(5.29)	(5.17)		(5.29)	(5.32)	(5.32)	(5.23)	(5.19)	(5.18)	(4.83)
LAL	212	NA	NA	245	245	NA	285	314	314	320	325	325	375
	(12.39			(12.95	(12.66)		(13.82)	(14.67)	(14.67)	(14.67)	(14.79)	(14.77)	(15.89)
Total mem-	1711	NA	NA	1892	1935	NA	2062	2141	2141	2181	2198	2200	2360
bership	(100.00)			(100.00)	(100.00)		(100.00)	(100.00)	(100.00)	(100.00)	(100.00)	(100.00)	(100.00)

Source: Various annual reports of KPACB.
Note: Figures in brackets are Percentage to total.

Particulars of Loans Issued

The distribution of the credit to different heads, like short-term, medium-term and long-term. The period-wise distribution of credit disbursed by co-operatives is shown in Table 5.11 and Table 5.11a.

Table 5.11
Particulars of Loan Issued by Thamaraipakkam PACB

(in Rupees)

Sl.No.	*Years*	*ST*	*MT*	*Non-agri purpose*	*Total*
	1	2	3	4	5
1.	1989-1990	891755	46983	367744	1306482
		(68.26)	(3.59)	(28.15)	(100.00)
2.	1990-1991	900792	18.252	358480	1277524
		(70.51)	(1.43)	(28.06)	(100.00)
3.	1991-1992	1274263	124560	166245	1565068
		(81.42)	(7.96)	(10.62)	(100.00)
4.	1992-1993	1432600	567929	621744	2622273
		(54.63)	(21.66)	(23.71)	(100.00)
5.	1993-1994	227456	136900	1081025	1445381
		(15.74)	(9.47)	(74.79)	(100.00)
6.	1994-1995	2906986	167,000	1590575	4664561
		(62.32)	(3.58)	(34.10)	(100.00)
7.	1995-1996	2913002	351252	2072450	5336704
		(54.59)	(6.58)	(38.83)	(100.00)
8.	1996-1997	3397402	342887	2505350	6245639
		(54.40)	(5.49)	(40.11)	(100.00)
9.	1997-1998	3833540	407400	4363633	8604573
		(44.55)	(4.74)	(50.71)	(100.00)
10.	1998-1999	4378456	530275	4186124	9094855
		(48.14)	(5.83)	(46.03)	(100.00)
11.	1999-2000	5852864	359005	5514070	11725939
		(49.91)	(3.06)	(47.03)	(100.00)
12.	2000-2001	7507306	2785475	4829359	15122140
		(49.64)	(18.42)	(31.94)	(100.00)
13.	2001-2002	7663676	840000	4734259	13237935
		(57.89)	(6.35)	(35.76)	(100.00)

Sources: Various Annual Reports of Thamaraipakkam PACB.
Note: Figures in brackets are percentage to total.

Table 5.11a
Particulars of Loan Issued by Kalathure PACB

(in Rupees)

Sl.No.	Years	ST	MT	Others	Total
	1	2	3	4	5
1.	1989-1990	429934	42210	NA	472144
		(91.00)	(8.94)		(100.00)
2.	1990-1991	NA	NA	NA	NA
3.	1991-1992	NA	NA	NA	NA
4.	1992-1993	217517	NA	33478	250995
		(86.66)		(13.34)	(100.00)
5.	1993-1994	647277	28783	200000	876060
		(73.88)	(3.29)	(22.83)	(100.00)
6.	1994-1995	NA	NA	NA	NA
7.	1995-1996	165039	46255	583000	794294
		(20.78)	(5.82)	(73.40)	(100.00)
8.	1996-1997	113335	10,000	35750	159085
		(71.24)	(6.29)	(22.47)	(100.00)
9.	1997-1998	236854	NA	NA	236854
		(100.00)			(100.00)
10.	1998-1999	664948	NA	NA	664948
		(100.00)			(100.00)
11.	1999-2000	325295	NA	208250	533545
		(60.97)		(39.03)	(100.00)
12.	2000-2001	30800	NA	61550	92350
		(33.35)		(66.65)	(100.00)
13.	2001-2002	2000325	NA	10500	2010825
		(99.48)		(0.52)	(100.00)

Source: Various Annual Reports of Kalathure PACB.
Note: Figures in brackets are percentage to total.

Short-term

Funds are needed for short-term (less than 15 months) for the purpose of purchasing seeds, fertilizers and pesticides. Short-term loans are normally, repaid after harvest of the crops. In the study the amount for short-term loan was maximum (81.42 percent) during 1991-92 and decreased to 57.89 percent during the period 2001-2002. It was absolutely increased in Kalathure PACBs (99.48 percent).

Medium-term Loan

A medium-term loan is required by farmers for the period of 3 to 5 years for the purpose of sinking of wells, the installation of tubewells, pumpsets, purchase of milk cows, etc. The medium-term loans issued during the period of study for agricultural purpose was minimum (1.43 per cent) during 1990-91 and maximum (21.66 per cent) during the period 1992-93 in Thamaraipakkam PACB. It was comparating less in Kalathure PACB (3.29 percent in 1993-94 and 8.94 percent in 1989-90.

Non-Agricultural Purposes

Co-operative banks are giving loans against pledging jewels. Though these loans cannot be said to be productive, they are very much needed by the lower and middle class people for meeting certain unavoidable expenses such as medical treatment, household expenses and for other non-agricultural purposes. Loan amount for non-agricultural purposes by 28.15 percent in the year 1989-1990 to 35.76 per cent in 2001-2002 in the case of Thamaraipakkam society. It was also increased to 13.34 per cent in 1992-1993, to 22.47 per cent in 1996-1997 in the society of Kalathure PACB (Vide table 5.11a).

Table 5.12 gives the information regarding the growth of deposit mobilisation in Thamaraipakkam PACB. PACBs receive their deposits from members and non-members. Fixed deposit is an important source of deposit for the co-operative banks. In the case of fixed deposit in Thamaraipakkam PACB was an appreciable upward trend from 1989-1990 to 2001-2002. The same table clearly depicted that during 1990-1991 and 1995-1996 it was in a dormant condition. It also shows that recurring deposits reached a steady increase over the period. Data are not available in case of Kalathure PACB.

Table 5.13 and 5.13a gives information regarding total members and actual loan users during 1989-1990 to 2001-2002 in Thamaraipakkam PACB and Kalathure PACB. The total members in Thamaraipakkam PACB, there is small change in the base year 1989-1990 to 2001-2002. This table also reflects the upward trend in the matter of actual loan users and number of persons with overdues from 1989-1990 to 2001-2002 in Thamaraipakkam PACB. Wilful default, heavy non-farm expenditure, failure of crops and old debts are the factors responsible for these overdues. Total

Table 5.12
Status of Savings by TPACB

(in Rupees)

Sl. No.	Years	Fixed Deposit	SB A/c	RD A/c
	1	2	3	4
1.	1989-1990	46550	267223	1845
		(100.00)	(100.00)	(100.00)
2.	1990-1991	8030	138835	2570
		(17.250)	(51.954)	(139.29)
3.	1991-1992	83500	31152	510.00
		(179.377)	(11.6576)	(27.642)
4.	1992-1993	1,07,530	6611.60	1240
		(230.998)	(2.474)	(67.208)
5.	1993-1994	303230	40582	135.00
		(651.407)	(15.186)	(7.317)
6.	1994-1995	664200	15016	3370
		(1426.852)	(46.783)	(182.655)
7.	1995-1996	144344	21837.85	11130
		(310.083)	(8.172)	(603.252)
8.	1996-1997	2088514	2850	12385
		(4486.60)	(10.646)	(671.273)
9.	1997-1998	3312860	471749	43765
		(7116.777)	(176.537)	(2372.086)
10.	1998-1999	4687671	169811.66	60255
		(10070.184)	(63.546)	(3265.853)
11.	1999-2000	6827414	473532	33572
		(14666.839)	(177.204)	(1819.620)
12.	2000-2001	7102211	189748	14696
		(15257.166)	(71.007)	(796.531)
13.	2001-2002	7249122	457472	10846
		(15572.764)	(171.194)	(587.859)

Source: Various Annual Reports of TPACB.
Note: Figures in brackets indicate indices.

members were increased in Kalathure PACB in the base year by 100.00 percent to 140.39 per cent in terminal year in 2001-2002. But actual loan user was declined trend in the period 1998-1999 than increased (Vide Table 5.13a).

The analysis shows that there has been a steady growth in the co-operative credit agencies both in Tamil Nadu and in Vellore district. However the effective members are on decline and the deposit mobilisation by PACBs are not satisfactory. A comparative study of Thamaraipakkam PACB and Kalathure PACB definitely proves the contention that Thamaraipakkam PACB has performed better than Kalathure PACB.

Table 5.13
Total Membership and Actual Loan Users - TPACB

Sl. No.	*Years*	*Total membership*	*Actual Loan user*	*Number of person for OD*
	1	2	3	4
1.	1989-1990	1519 (100.00)	406 (100.00)	170 (100.00)
2.	1990-1991	1555 (102.369)	520 (128.078)	182 (107.058)
3.	1991-1992	808 (53.192)	284 (69.950)	148 (87.058)
4.	1992-1993	1025 (67.478)	320 (78.817)	165 (97.058)
5.	1993-1994	1116 (73.469)	595 (146.551)	128 (75.294)
6.	1994-1995	1223 (80.513)	814 (200.492)	134 (78.823)
7.	1995-1996	1284 (84.529)	861 (212.068)	81 (47.647)
8.	1996-1997	1464 (96.379)	1286 (316.748)	95 (55.882)
9.	1997-1998	1477 (97.235)	1047 (257.881)	170 (100.00)
10.	1998-1999	1530 (100.724)	1487 (366.256)	238 (140.00)
11.	1999-2000	1688 (111.125)	1099 (270.689)	208 (122.352)
12.	2000-2001	1576 (103.752)	1141 (281.034)	264 (155.294)
13.	2001-2002	1597 (105.134)	1018 (250.738)	343 (201.764)

Source: Various Annual Reports of TPACB
Note: Figures in brackets indicate indices

Table 5.13a
Total Membership and Actual Loan Users in KPACB

Sl. No.	*Years*	*Total membership*	*Actual Loan user*	*Number of person for OD*
	1	2	3	4
1.	1989-1990	1681 (100.00)	97 (100.00)	299 (100.00)
2.	1990-1991	NA	NA	NA
3.	1991-1992	NA	NA	NA
4.	1992-1993	1895 (112.730)	37 (38.144)	221 (73.913)
5.	1993-1994	1935 (115.110)	101 (104.123)	145 (48.494)
6.	1994-1995	NA	NA	NA
7.	1995-1996	2062 (122.665)	231 (238.144)	180 (60.200)
8.	1996-1997	2181 (129.744)	231 (238.144)	233 (77.926)
9.	1997-1998	2141 (127.364)	43 (44.329)	271 (90.635)
10.	1998-1999	2181 (129.744)	82 (84.536)	300 (100.334)
11.	1999-2000	2198 (130.755)	365 (376.288)	168 (56.187)
12.	2000-2001	2200 (130.874)	368 (379.381)	225 (75.250)
13.	2001-2002	2360 (140.392)	399 (411.340)	120 (40.133)

Source: Various Annual Reports of KPACB.
Note: Figures in brackets indicate indices.

REFERENCES

1. Palaniappa Mudaliar, A, *The Co-operative Movement in Madras State-A Brief Survey,* The Tamil Nadu Co-operative Union, Madras, 1958, p. 1.
2. Victor, M.A., *Co-operation in Madras State,* Palaniappa Brothers, Madras, 1964, p. 2, 3.
3. Thirunarayanan, R., *Co-operative Banking in India,* Mittal Publications, New Delhi, 1996, P. 31, 35.
4. Ibid. p. 36.
5. Tamil Nadu Journal of Co-operation, Chennai, September - 2002, p. 11.
6. Annual Reports, Vellore District Central Co-operative Bank 1998-99.

Chapter - 6

SOCIO-ECONOMIC CHARACTERISTICS OF SAMPLE RESPONDENTS: ANALYSIS OF THE PRIMARY DATA

INTRODUCTION

This chapter deals with the interpretation of the primary data collected from 371 sample respondents of two PACBs, for the agricultural year 2000-2001. The most common socio-economic characteristics such as sex, age, social status and educational level are discussed. A few more variables such as occupation, annual income, sources of irrigation, land holding and flow of credit etc, which are likely to have an influence on the loan receiving and repaying capacity of the sample members are also added in this analysis. The analysis based on frequency tables for above said socio-economic variables are discussed in this chapter.

Table 6.1 shows that among the 371 sample respondents chosen for the study. 218 are from Kalathure PACB and 153 are from Thamaraipakkam PACB. Since 10 percentage of the total membership has been chosen at random on proportionate sampling basis, the numbers differ for each sample village.

From the given table it be said that small and marginal farmers farm the bulk of the membership of both PACBs. Again, it can also be noted that those who are holding land between 1 and 4 hectares are larger in Thamaraipakkam compared to Kalathure where those who own below 1 hectare more in number. This may be a reason for more defaulters at Kalathure. It is so because the small and marginal farmers do not have the capacity to repay the loan. Landlessness seems to be more among the members of Kalathure.

Table 6.1
Land Holdings of Sample Respondents

Sl. No.	Land holding (in hectare)	Thamaraipakkam PACB	Kalathure PACB	Total
	1	2	3	4
1.	Below 1 hectare	40 (26.14)	65 (29.82)	105 (28.30)
2.	1 - 2	45 (29.41)	41 (18.81)	86 (23.18)
3.	2 - 4	27 (17.65)	22 (10.07)	49 (13.21)
4.	4 - 8	8 (5.23)	17 (7.80)	25 (6.74)
5.	8 and above	1 (0.65)	3 (1.38)	4 (1.08)
6.	Agriculture labourers	24 (15.69)	27 (12.39)	51 (13.75)
7.	Rural Artisan	2 (1.31)	11 (5.05)	13 (3.50)
8.	Others (landless agri. labourers)	6 (3.92)	32 (14.68)	38 (10.24)
	Total	153 (100.00)	218 (100.00)	371 (100.00)

Source: Field data
Figures in parentheses are percentages.

Table 6.2
Sex-wise Distribution of Sample Respondents

Sl. No.	Name of the Sex	Thamaraipakkam PACB	Kalathure PACB	Total
	1	2	3	4
1.	Male	148 (96.73)	209 (95.87)	357 (96.23)
2.	Female	5 (3.27)	9 (4.13)	14 (3.77)
	Total	153 (100.00)	218 (100.00)	371 (100.00)

Source: Field data.
Figures in parentheses are percentages.

Table 6.2 sex-wise classification of the respondents has shown that male got a high representation in the total sample. There are 357 male respondents (96.23 percent) and 14 (3.77 percent) female respondents. One can infer from this that there are only a few land owner (pattadar) among women.

Table 6.3 depicts the agewise distribution of the respondents. Majority of the sample respondents are from the age group of 36-59 years. It accounts 64.15 percent of the sample. Percentage of respondents which come under the group of 35 years and that of more than 60 years accounts 18.87 per cent and 16.98 per cent respectively.

Table 6.3
Age-wise Distribution of the Sample Respondents

Sl. No.	Age-group	Thamaraipakkam PACB	Kalathure PACB	Total
	1	2	3	4
1.	Below 35	31 (20.26)	39 (17.89)	70 (18.87)
2.	36 - 59	100 (65.36)	138 (63.30)	238 (64.15)
3.	60 and above	22 (14.38)	41 (18.81)	63 (16.98)
	Total	153 (100.00)	218 (100.00)	371 (100.00)

Source: Field data.
Figures in parentheses are percentages.

Table 6.4
Community-wise Distribution of Sample Respondents

Sl. No.	Name of the Community	Thamaraipakkam PACB	Kalathure PACB	Total
	1	2	3	4
1.	SC/ST	31 (20.26)	50 (22.94)	81 (21.83)
2.	BC	58 (37.91)	34 (15.60)	92 (24.80)
3.	MBC	48 (31.37)	130 (59.63)	178 (47.98)
4.	OC	16 (10.46)	4 (1.83)	20 (5.39)
	Total	153 (100.00)	218 (100.00)	371 (100.00)

Source: Field data
Figures in parentheses are percentages.

It is clearly indicated in the Table 6.4, that most of the respondents belongs to MBC (47.98 percent), followed by BC (24.80 percent), SC/ST (21.83 percent) and OC (5.39 percent). These data do not say that MBCs are dominating and OCs are a few in the PACB, but this represents communal combination of castes in the sample villages.

Education is the important instrument of social inputs. The educational status would have a profound influence in the financial management of farm and home and, in turn, on the need for credit.

Table 6.5 shows that educational status of the respondents. Respondents who are having more than 10 years of schooling represents 8.62 percent of the total sample. Majority of the respondents studied upto 8th standard Illiterate constitute 22.91 percent of the total respondents. The percentage of illiterate members are more in Kalathure PACB.

Table 6.5
Educational Status of the Respondent

Sl. No.	*Educational Status*	*Thamaraipakkam PACB*	*Kalathure PACB*	*Total*
	1	2	3	4
1.	Illiterate	26 (16.99)	59 (27.06)	85 (22.91)
2.	1-8 std	81 (52.94)	115 (52.75)	196 (52.83)
3.	9 std and 10th	33 (21.57)	25 (11.47)	58 (15.63)
4.	HSC	5 (3.27)	7 (3.21)	12 (3.23)
5.	Degree and others	8 (5.23)	12 (5.50)	20 (5.39)
	Total	153 (100.00)	218 (100.00)	371 (100.00)

Source: Field data.
Figures in parentheses are percentages.

Table 6.6
Occupational Status of the Respondents

Sl. No.	*Occupation*	*Thamaraipakkam PACB*	*Kalathure PACB*	*Total*
	1	2	3	4
1.	Agriculture	75 (49.02)	137 (62.85)	212 (57.14)
2.	Agriculture & Allied	27 (17.65)	30 (13.76)	57 (15.36)
3.	Agriculture & Business	32 (20.92)	24 (11.01)	56 (15.09)
4.	Employment in outside Agriculture	18 (11.76)	16 (7.34)	34 (9.16)
5.	Rural Artisan	1 (0.65)	11 (5.05)	12 (3.23)
	Total	153 (100.00)	218 (100.00)	371 (100.00)

Source: Field data.
Figures in parentheses are percentages.

Table 6.6 deals with the occupation-wise classification of the respondents. Agriculture is the major occupation in the study area. Majority of the sample respondents were depending on agriculture as their occupation i.e., 57.14 percent. Around 15.00 per cent each were engaged in agriculture and allied activities and Business with agriculture respectively. The table clearly bringsout the fact that less number of respondents depends wholly on agriculture in Thamaraipakkam compared to Kalathure. More than 51 percent of the respondents supplements their income from with income earned from outside agriculture or from allied activities. It can be said that more cash flow will be there among the respondents of Thamaraipakkam and hence they can easily repay the credit borrowed from PACB where as in case of Kalathure outside cash inflow is much less.

Similarly, rural artisans are living more in Kalathure (5.05 percent) compared to Thamaraipakkam (0.65 percent only). Which may also affect their repaying capacity.

Table 6.7
Annual Income of the Respondents Family

Sl. No	*Income Classification*	*Thamaraipakkam PACB*	*Kalathure PACB*	*Total*
	1	2	3	4
1.	Below Rs.10000	12 (7.84)	40 (18.35)	52 (14.02)
2.	Rs.10001-Rs.15000	39 (25.49)	59 (27.06)	98 (26.42)
3.	Rs.15001-Rs.20,000	40 (26.14)	57 (26.15)	97 (26.15)
4.	Rs.20,001-Rs.30,000	36 (23.53)	39 (17.89)	75 (20.22)
5.	Rs.30001 and above	26 (16.99)	23 (10.55)	49 (13.21)
	Total	153 (100.00)	218 (100.00)	371 (100.00)

Source: Field data.
Figures in parentheses are percentages.

Annual income of the respondent family is given in Table 6.7 clear picture. In the total sample 14.02 percent are having only less than Rs.10,000 of annual income. Respondents who are having more than Rs.30,001 as annual income constitute 13.21 percent of the total sample. Picture is different between societies. In Thamaraipakkam society only 7.84 percent having less than Rs.10,000 whereas 18.35 per cent of the sample at the Kalathure society are having less than Rs.10,000 as annual income. Same time 40.52 percent of the respondents in the Thamaraipakkam society having more than Rs.20,000 of annual income whereas only 28.44 percent of the Kalathure society having more than Rs.20,000 of annual income.

The inference that can made from the above discussion is that members belonging to Thamaraipakkam society are keeping high income profile compared to the members in the Kalathure society.

Table 6.8 clearly mentions the status of operational land holding among sample households. Majority of the sample are holding own land i.e., 92.81 per cent. In Thamaraipakkam 94.33 percent of the PACB beneficiaries are holding own land. In the case of Kalathure PACB beneficiaries 91.62 per cent are holding own land. Lease in and or lease out constitute 5.67 percent of the sample in Thamaraipakkam whereas it is 8.38 percent in Kalathure. Lease in and out are more in Kalathure compared to Thamaraipakkam area.

Table 6.8
Operational Land Holdings

Sl. No.	Land Status	Thamaraipakkam PACB	Kalathure PACB	Total
	1	2	3	4
1.	Own land	133 (94.33)	164 (91.62)	297 (92.81)
2.	Lease in & out	8 (5.67)	15 (8.38)	23 (7.19)
	Total	141 (100.00)	179 (100.00)	320 (100.00)

Source: Field data.
Figures in parentheses are percentages.

Table 6.9 indicates the operational land holding of the respondents. There are households which are having both wet and dry land holdings. In the total sample, 233 are keeping wet land but only 186 of them are cultivating the land. The difference between ownership land holding and operational land holdings are relatively high in Thamaraipakkam. Though 103 of the sample households are having own land only 74 of them are cultivating it in Thamaraipakkam. But in Kalathure 112 of the sample household out of 130 are cultivating these wet lands.

Table 6.9
Total Land & Cultivation

Sl. No.	Land Particular	Thamaraipakkam PACB		Kalathure PACB		Total	
		Count	Mean	Count	Mean	Count	Mean
	1	2	3	4	5	6	7
1.	Total wet Land Holder	103	3.25	130	3.52	233	3.40
2.	Wet cultivator only	74	3.22	112	3.02	186	3.10
3.	Total Dry Land holder	80	2.48	91	2.74	171	2.62
4.	Dry Cultivator Only	35	2.16	59	2.58	94	2.43

Source: Field data.
Figures in parentheses are percentages.

In the case of dry land, 94 samples out of 171 are cultivating. Hence the gap between number of landholdings and number of operational land holdings are higher in the case of Thamaraipakkam.

Table 6.10
Sources of Irrigation

Sl. No.	*Particulars of Irrigations*	*Thamaraipakkam PACB*	*Kalathure PACB*	*Total*
	1	2	3	4
1.	Tank	21 (16.28)	49 (32.45)	70 (25.00)
2.	Well with Pump-Sets	96 (74.42)	99 (65.56)	195 (69.64)
3.	Well with oil-Engine	12 (9.30)	3 (1.99)	15 (5.36)
	Total	129 (100.00)	151 (100.00)	280 (100.00)

Source: Field data.
Figures in parentheses are percentages.

Water is an important input for agriculture and without this, it cannot be possible for cultivation. Details of sources of irrigation are presented in the Table No.6.10. A total of 69.64 percent of the sample beneficiaries are having well with pump-set as their source of irrigation. It has been more in Thamaraipakkam (74.42 per cent) compared to kalathure (65.56 per cent) area. The second sources of irrigation constitute Tank irrigation i.e 25 percent. It has been more in kalathure area (32.45 per cent) compared to Thamaraipakkam (16.28 per cent) area and 5.36 percent of the beneficiaries were depends on well with oil-engine as their source of irrigation. The percentage of sample households depending upon uncertain Tank water irrigation is more at Kalathure. This perhaps can explain the production uncertainty at Kalathure and hence the resultant behaviour of putting the co-operative society in to a bad category.

Table 6.11 clearly indicates the particulars of savings of the sample households. Sources of savings can be grouped into PACB, commercial bank, LIC, post office and others (i.e. included Peerless, Relatives & Friends). Among the total sample households, 62.80 per cent have no savings at all. They are more among sample households of Kalathure (72.94 per cent) than among Thamaraipakkam, sample households (48.37 per cent). This is due to the fact that the numbers at Kalathure PACB are more agricultural labourers having less landed property.

Table 6.11
Mode of Savings

Sl. No.	Name of the Institution	Thamaraipakkam PACB	Kalathure PACB	Total
	1	2	3	4
1.	PACB	29 (36.71)	5 (8.47)	34 (24.64)
2.	Commercial Bank	5 (6.33)	17 (28.82)	22 (15.94)
3.	LIC + Post Office	40 (50.63)	22 (37.29)	62 (44.93)
4.	Others*	5 (6.33)	15 (25.42)	20 (14.49)
	Total Savers	79 (100.00)	59 (100.00)	138 (100.00)
	Total Non-Savers	74	159	233
	Grand Total	153	218	371

Source: Field data
Figures in parentheses are percentages.
*Others : Refers to relatives, friends, chit funds, Peerless, insurance etc.

PACB plays only lesser role to mobilise the rural savings, i.e., only 24.64 percent of the sample used PACB for their saving purpose. Thamaraipakkam PACB shows a better performance in this regard (18.95 percent) where as it is only 2.29 percent of the samples from Kalathure PACB. The poor working conditions of PACB at Kalathure might have discouraged the people to deposit their title savings at PACB.

The analysis indicates that savings habit is much less among the sample households. This may be because of this poor income level which leaves nothing to save. The PACB's have not mobilised even the little savings habit that is found among the members of PACBs.

Table 6.12 gives the information about the amount of savings of the respondents. Majority of the sample (56.52 per cent) have

Table 6.12
Amount of Savings

Sl. No.	Amount Classification	Thamaraipakkam PACB	Kalathure PACB	Total
	1	2	3	4
1.	Below Rs.5000/=	44 (55.70)	34 (57.63)	78 (56.52)
2.	Rs. 5001-Rs.25,000/=	28 (35.44)	14 (23.73)	42 (30.44)
3.	Rs.25,001 and above	7 (8.86)	11 (18.64)	18 (13.04)
	Total	79 (100.00)	59 (100.00)	138 (100.00)

Source: Field data.
Figures in parentheses are percentages.

only less than Rs.5000 as their total annual savings. Percentage of sample having more than Rs.25,001 as saving is relatively higher in Kalathure than in Thamaraipakkam. This is may be due to more unequal land holding structure at Kalathure.

It has been clearly mentioned in the Table 6.13, that 198 (53.37 per cent) out of 371 samples received loan from PACB during 2000-2001. More members in the Thamaraipakkam society are getting loans than the members in Kalathure society.

Table 6.13
Particulars of Loan Availed During 2000-01

Sl. No.	*Items*	*Thamaraipakkam PACB*	*Kalathure PACB*	*Total*
	1	2	3	4
1.	Loan availed	123 (80.39)	75 (34.40)	198 (53.37)
2.	Loan not availed	30 (19.61)	143 (65.60)	173 (46.63)
	Total	153 (100.00)	218 (100.00)	371 (100.00)

Source: Field data.
Figures in parentheses are percentages.

A lesser proportion has got credit facility from PACB of Kalathure because more number of sample households have not paid their dues to the Bank in time and hence they could not re-borrow.

Table 6.14 shows that there are some reasons for sample households not approaching any loan from PACB. Out of 371 sample 173 respondents have not received any kind of loan for their agricultural purpose. It reasons given by the respondents at the time of data collection are considered, 31.21 percent of the respondents have not approached because of previous dues and second major reason is 'JAPTI' which is 26.01 percentage level. It can be seen that this is only at Kalathure society. Third person surety problem form also 11.56 percent. This problem is more at Thamarapakkam than at Kalathure.

Table 6.15 shows the picture of the purpose for which the loan is sought by the respondents. It is made clear that TPACB has got good performance than KPACB in the issue of loans. Among the total, 64.65 percent have received the loan for purchasing fertilisers. Others have received the loan for purchasing consumer items (14.65%), cattle (11.11%) and agricultural implements (4.04%).

Table 6.14
Reasons for not Receiving any Loan from Society

Sl. No.	*Reasons*	*Thamaraipakkam PACB*	*Kalathure PACB*	*Total*
	1	2	3	4
1.	Previous dues	10 (33.33)	44 (30.77)	54 (31.21)
2.	No cultivation for water scarcity	5 (16.67)	12 (8.39)	17 (9.83)
3.	Surety problem	8 (26.67)	12 (8.39)	20 (11.56)
4.	Unable to repay	4 (13.33)	4 (2.80)	8 (4.63)
5.	Recommendation & Commissions	-	3 (2.10)	3 (1.74)
6.	Own money	1 (3.33)	18 (12.59)	19 (10.98)
7.	Past *Japti* due to default	-	45 (31.47)	45 (26.01)
8.	Loan received from *Mandi* (private)	2(6.67)	5 (3.49)	7 (4.04)
	Total	30(100.00)	143 (100.00)	173 (100.00)

Source: Field data.
Figures in parentheses are percentages.

Table 6.15
Purpose of Loan Sought from PACB

Sl. No.	*Purposes*	*Thamaraipakkam PACB*	*Kalathure PACB*	*Total*
	1	2	3	4
1.	Fertilizers	63 (51.22)	65 (86.67)	128 (64.65)
2.	Cattle	20 (16.26)	2 (2.67)	22 (11.11)
3.	Agri Implements (Motor-Pump-Sets)	7 (5.69)	1 (1.33)	8 (4.04)
4.	Consumer Loan (TV, TVS-50, Cycle)	29 (23.53)	-	29 (14.65)
5.	Jewell Loan*	4 (3.25)	7 (9.33)	11 (5.55)
	Total	123 (100.00)	75 (100.00)	198 (100.00)

Source: Field data.
Figures in parentheses are percentages
* This loan can be spent on any purpose

Though jewell loans go under the name of 'agricultural operations', mostly it is to meet the exigencies such as expenditure on health, education, social functions etc.

Table 6.16
Amount of Loan

Sl.No.	*Name of the PACBs*	*Count*	*Mean*
	1	2	3
1.	Thamaraipakkam PACB	123	16444
2.	Kalathure PACB	75	7700
	Total	198	13132

Source: Field data.
Figures in parentheses are percentages.

Performance of a PACB can be measured by considering the number of loan issued and amount of loan disbursed. Thamaraipakkam PACB has issued loans for 123 samples i.e., 80.39 percent of the samples. At the same time only 75 or 34.40 percent of the samples belonging to Kalathure PACB have received loan. There is a big gap between Thamaraipakkam and Kalathure with respect the mean loan amount. In Thamaraipakkam PACB mean loan amount is Rs.16,444, but it is just Rs.7700 in Kalathure PACB (Table 6.16).

Table 6.17
Alternative Sources of Finance

Sl. No	*Sources*	*Thamaraipakkam PACB*	*Kalathure PACB*	*Total*
	1	2	3	4
a.	Institutional			
1.	Commercial Bank	13 (16.88)	22 (19.47)	35 (18.42)
2.	RRB & LD Bank	1 (1.30)	9 (7.96)	10 (5.26)
B.	Non-Institutions			
3.	Money Lenders/ Commission Agent/ *mandis**	37 (48.05)	70 (61.95)	107 (56.32)
4.	Relatives & friends	26 (33.77)	12 (10.62)	38 (20.00)
	Total	77 (100.00)	113 (100.00)	190 (100.00)

Source: Field data.
Figures in parentheses are percentages.
* Mandis : Trading houses who finance the agricultural operations and will buy the harvest.

The credit requirements of the rural masses are met by different sources viz: institutional as well as non-institutional. Institutional source include Government, Commercial Banks, Regional Rural Banks and Co-operatives. While non-institutional agencies comprise of money-lenders both agricultural and professional, commission agents (mundis), friends and relatives and others. The details of the credit agencies in respect of sample households are presented in Table No.17. Out of 371 sample size, 198 respondents received loan from PACB.

Table 6.17 indicates that among those samples that are taken loan from PACB, 190 (95.95%) are dissatisfied with the loan amount. Most of them (76.32%) have gone for non-institutional Finance for their requirements. In Thamaraipakkam PACB, it comes upto 81.82 percent and it is 72.57 percent in Kalathure PACB Area. The growing cost of cultivation has pushed the farmers to go for any other alternative sources of loan inspite of the higher interest rate.

Table 6.18
Subsidy from PACBs for Agricultural Activities 2000-2001

Sl. No.	*Purpose of subsidy*	*Thamaraipakkam PACB*	*Kalathure PACB*	*Total*
	1	2	3	4
1.	Fertilizers (Seeds)	4 (17.39)	2 (50.00)	6 (22.23)
2.	Agri Implements	1 (4.35)	-	1 (3.70)
3.	Motor-Pump-sets	1 (4.35)	-	1 (3.70)
4.	Cattle (IRDP)	17 (73.91)	2 (50.00)	19 (70.37)
	Total	23 (100.00)	4 (100.00)	27 (100.00)

Source: Field data.
Figures in parentheses are percentages.

Twenty seven respondents out of total sample have received some kind of subsidy loan. One can observe that it is 23 respondents in Thamaraipakkam PACB, and just 4 in Kalathure PACB. Most of them got subsidy, when they have taken loan for buying cattle. It can be inferred from this table that more from Thamaraipakkam have attempted to enhance their income by going for agricultural allied activities.

Table 6.19 clearly sums up the reasons for people approaching to PACB for financial support for agricultural activities. Respondents have been asked to point out the reasons for which they have selected PACBs for taking loan. Around 39.39 percent of them pointed out a combination of low interest rate, Easy terms of repayment and near to residence. And another 25.76 percent is attracted because of low interest rate prevailing in PACB loans. The reasons given by them indicate that loans from PACBs are soft loans and hence people prefer it.

Table 6.19
Reasons for Approaching PACB for Financial Support for Agricultural Activities

Sl. No.	*Reasons*	*Thamaraipakkam PACB*	*Kalathure PACB*	*Total*
	1	2	3	4
1.	Low Interest Rate	15 (12.19)	36 (48.00)	51(25.76)
2.	Easy terms of Repayment	13 (10.57)	2 (2.67)	15.(7.58)
3.	Near to Residence	5 (4.07)	18 (24.00)	23 (11.62)
4.	To enjoy subsidy	5 (4.07)	-	5 (2.53)
5.	To avoid Risk	5 (4.07)	1 (1.33)	6 (3.03)
6.	To avail consultancy/ guidancy	1 (0.81)	-	1 (0.50)
7.	Low Interest Rate/Easy terms of Repayment/Near to residence	61 (49.59)	17 (22.67)	78 (39.39)
8.	Easy terms of Repayment/ Near to Residence	6 (4.88)	1 (1.33)	7 (3.53)
9.	Low Interest Rate/Easy terms of Repayment/to enjoy subsidy	11 (8.94)	-	11 (5.56)
10.	Immediate disbursement	1 (0.81)	-	1 (0.50)
	Total	123 (100.00)	75 (100.00)	198 (100.00)

Ninenty seven out of 198 have expressed that they do face certain problems in getting the loan from PACB. Respondents have pointed out various problems that they have faced at the time of getting the loan. Respondents are affected by various problems. viz; deposit, third person surety, delay in sanctioning the loan, political involvement, and others. Among them surety is the single problem that many (39.18%) of the respondents have pointed out. And another 29.89 percent are severely affected by the condition laid down by PACB that they have to deposit same amount from the loan amount. Delay in sanctioning of the loan is another single problem pointed out by many (11.34%). The table 5.20 presents the problems of respondents is getting loan from PACB.

Table 6.20
Problem Faced by Respondents, at the Time of Receiving Loan from PACB

Sl. No.	*Name of the problems*	*Thamaraipakkam PACB*	*Kalathure PACB*	*Total*
	1	2	3	4
1.	To deposit	-	1 (1.64)	1 (1.03)
2.	Surety	27(75.00)	11 (18.03)	38 (39.18)
3.	Delay in sanctioning the loan	1 (2.78)	10 (16.39)	11 (11.34)
4.	Political involvement	-	2 (3.28)	2 (2.56)
5.	Commission asked by staff	-	7 (11.48)	7 (7.22)
6.	To deposit/surety/delay sanctioning the loan	7 (19.44)	22 (36.07)	29 (29.89)
7.	Surety/commission	1 (2.78)	5 (8.19)	6 (6.19)
8.	Delay sanctioning the loan/ political involvement/ commission	-	3 (4.92)	3 (3.09)
	Total	36 (100.00)	61 (100.00)	97 (100.00)

Source: Field data.
Figures in parentheses are percentages.

Table 6.21
Reasons for Default

Sl. No.	*Reasons*	*Thamaraipakkam PACB*	*Kalathure PACB*	*Total*
	1	2	3	4
1.	Lack of Suitable / adequate staff for the purpose of efficient supervision and recovery of loan	3 (21.42)	1 (4.55)	4 (11.11)
2.	Clearing old debt to money lenders/commission Agent/ Relatives	2 (14.29)	4 (18.18)	6 (16.66)
3.	Expenditure on consumption	7 (50.00)	6 (27.27)	13 (36.11)
4.	Unfavourable price structure	-	2 (9.09)	2 (5.56)
5.	Late sugarcane cutting late bill sanctioning	-	2 (9.09)	2 (5.56)
6.	Natural calamities for drought/ floods	-	2 (9.09)	2 (5.56)
7.	Low yield per acre	2 (14.29)	5 (22.73)	7 (19.44)
	Total	14 (100.00)	22 (100.00)	36 (100.00)

Defaulters are more in number among the sample households at Kalathure compared to Thamaraipakkam. It is worth to note that low yield from their crops also make them to fail from repaying the credit. Table No.6.21 gives an idea about the default. Many of the (36.11%) respondents used the loan amount for consumption purpose. Because of its unproductive nature, they can't repay the loan amount within the stipulated period. And another 16.66 percent loanees utilised the loan amount for the repaying old debt from other non-organisational sources where interest rates are high. Uncertain climatic conditions and floating agricultural prices made some of the respondents inefficient to repay the loan.

Table 6.22
Suggestions Given by Beneficiaries

Sl. No.	*Suggestions*	*TPACB*	*KPACB*	*Total*
	1	2	3	4
1.	No suggestions	33 (21.57)	72 (33.03)	105 (28.30)
2.	Good society-Good Management	55 (35.95)	10 (4.59)	65 (17.52)
3.	Bad society -Bad Management	4 (2.61)	81 (37.15)	85 (22.91)
4.	Third person surety problem	13 (8.49)	6 (2.75)	19 (5.12)
5.	No Transport facility	3 (1.96)	9 (4.13)	12 (3.23)
6.	Farmers will expect loan for Housing, van, car, Auto	6 (3.92)	-	6 (1.62)
7.	Sources will available by PACB - supply of seeds - Warehousing facilities - sales facilities	10 (6.54)	-	10 (2.69)
8.	Loan amount inadequate must increased	21 (13.73)	8 (3.67)	29 (7.82)
9.	No consumer loan	2 (1.31)	24 (11.01)	26 (7.01)
10.	Two Board members must from below poverty group	-	1 (0.46)	1 (0.27
11.	Need not surety for green card holders	3 (1.96)	-	3 (0.81)
12.	Selection of PACB staff through by recruitment	3 (1.96)	6 (2.75)	9 (2.43)
13.	Repayment period should be extend	-	1 (.46)	1 (0.27)
	Total	153 (100.00)	218 (100)	371 (100.00)

Source: Field data.
Figures in parentheses are percentages.

Table 6.22 depicts that the respondents suggestions to improve the working of PACBs. It is recognised that there are some loopholes in the credit process and in the working of PACB. Many farmers could not come forward to discuss the measures to improve the working of PACB because of fear for the officials. Around by 28.30 percent respondents have given no suggestion. About 22.91% respondents have said that PACB are not functioning well and are managed badly. It is much clear that 37.15 percent respondents of Kalathure society have said that the society is bad compared to get 2.61 percent Thamaraipakkam society.

It is necessary for the farmers to furnish third party surety to get loan and (5.12%) have complained that this is a big problem. This often involves furnishing of mutual surety and this result in undue burden on both the recipient as well as the guarantor. It one fails to repay the other cannot get loan even if there is no due for him. With PACB also has resulted in undue delay in disbursement of loans, which took away the benefits that otherwise would accrue if timely agricultural operation had taken place.

Many of the farmers expressed the suggestion (2.43%) that PACB staff should be based on good selection and training and this will help to avoid certain drawbacks.

The entire discussion points out that the effective members who presently use the loan facilities are not satisfactory if total membership is taken into consideration. Further it is also a fact that there is a widening 'credit gap' is loan amount needed for agricultural operations and the amount sanctioned by PACBs. This credit gaps push the people to take recourse to non-formal credit agencies. Another factor to be noted is that PACB's have done very little in mobilising the rural savings. The problem of overdues are on its increase. Most of the borrowers have approached the PACBs for the low interest charged and the subsidy available. If these facilities are with drawn in effective membership may fall.

Chapter - 7

AN ANALYSIS OF WORKING OF SAMPLE PACBs (HYPOTHESES TESTING)

INTRODUCTION

This chapter has been devoted to the analysis of primary data. Hypotheses have been tested and conclusions have been drawn. The chapter has been arranged into three parts : in the first part all the hypotheses have been tested taking all the samples together. In the second part, the hypotheses have been tested for both the sample societies individually. In the third part statistical treatment has been made on the secondary data.

PART I - (BOTH SOCIETIES)

The calculated chi-square value is 74.49 and p value is .000, which is significant at 1 percentage level and hence one can reject the null-hypothesis.

There is a relationship between size of credit and size of the land holding. Correlation co-efficient is 0.3059, which indicates positive (direct) relationship between size of credit and size of land holding. So the research hypothesis "size of credit is directly related to size of farm" can be accepted.

Hypothesis-I: Size of credit is directly related to size of farm

H_0 : There is no relationship between size of credit and size of the land holding.

Table 7.1 denotes that among landless 89.5 per cent are not receiving any loan and even a single member is not receiving a loan amount of more than Rs.10,001. But considering those samples, which are having more than 8 Hectare of land one can observe that 17.2 percent do not receive any loan and 48.3 percent have received loan amount more than Rs.10,001/-.

Table 7.1
Size of Land and Size of Credit

Sl. No.	Loan Amount → ↓ Land Group	No loan	Upto Rs.5000	Rs.5001- Rs.10,000	Rs.10,001- and above	Total	Chi-square	P value
	1	2	3	4	5	6		
1.	No land	34 (89.5)	2 (5.3)	2 (5.3)	0	38 (100.00)	74.49	0.000**
2.	Upto 1 hectare	59 (57.3)	21 (20.4)	12 (11.7)	11 (10.7)	103 (100.00)		
3.	1-2	29 (48.3)	15 (25.0)	9 (15.0)	7 (11.7)	60 (100.00)		
4.	2-4	31 (36.9)	17 (20.2)	18 (21.4)	18 (21.4)	84 (100.00)		
5.	4-8	15 (26.3)	12 (21.1)	16 (28.1)	14 (24.6)	57 (100.00)		
6.	8 and above	5 (17.2)	4 (13.8)	6 (20.7)	14 (48.3)	29 (100.00)		
	Total	173 (46.6)	71 (19.1)	63 (17.0)	64 (17.3)	371 (100.00)		

Source: Primary data computed.
Note: Figures in brackets are percentage to row total.

If one considers both the landless and small land holdings of less than 1 hectare 65.95 percent of them do not receive any loan. But the same for those land holding of more than 4 Hectare, 23.25 percentage do not receive any loan. Among landless and small land holdings of less than 1 Hectare, 80 percentage were received loan amount of Rs.10,000/-. But 32.55 percentage of sample having more than 4 Hectare of land received loan amount of more than Rs.10,000/-

Hypothesis-II: The Co-operative Credit has Not Grown to the Needs of Farmers

H_0: There is no relationship between satisfaction of loan takers and their loan amount.

Table 7.2 gives a clear picture of the credit gap. Credit gap is estimated as the difference between the level of borrowing (actual loan amount given by PACB) and the actual amount spent by the borrower for the successful implementation of the agricultural activities. It has been calculated by considering the difference between the loan received for cultivating 1 acre of paddy and the actual cost of cultivation as per the respondent. PACBs are giving Rs.5280 as loan amount for each acre of land for cultivation of paddy. There may be differences in the cost of cultivation among respondents because of their differences in the consumption of agricultural inputs. Table 7.2 reports that per acre cost of cultivation reported by the loan takers in the loan group of below Rs.5000 is Rs. 5367. The cost of cultivation is Rs. 6790 and Rs. 6698 respectively for loan takers in the range of Rs. 5001-Rs. 10,000 and Rs. 10,001 and above, which makes the credit gap ununiform. That means, they have to spend more than the bank loan for the successful completion of the agricultural activities.

Inadequate loan amounts make the cultivators to compromise on their expenditure on agricultural inputs like HY seeds, fertilizers, pesticides and irrigation, which leads to a production far below the optimum level and make agriculture an unremunerative enterprise. Consequently the low production leads to defaults and overdues. It can be noticed here that the researcher has given data only for paddy and for other commercial crops, such as banana, coconuts, sugarcane etc., the credit gap may be higher.

Table7.2

Estimation of Credit Gap based on the Cost of Paddy Cultivation

Sl. No.	*Loan amount*	*Actual loan amount per acre*	*Cost of cultivation per acre*	*Credit gap difference*
	1	2	3	4
1.	Below 5000	Rs.5280	Rs.5367	Rs. 87
2.	Rs. 5001-Rs. 10,000	Rs.5280	Rs.6790	Rs.1510
3.	Rs. 10,001 and above	Rs.5280	Rs.6698	Rs.1418

Source: Field data.

The analysis given here shows that the calculated chi-square value is 9.023. As the p value is 0.01 which is highly significant at 1 percentage level, the null-hypothesis can be rejected. There is a relationship between satisfaction of loan takers and their loan amount.

Among the respondents who have taken loan amounts of less than Rs. 5000/-, 63.4 percent are not satisfied with the loan they received. But majority of the respondents who are receiving more than Rs. 10,000/- are satisfied. Considering the total loan takers one can observe that 101 (51 per cent) out of 198 are not satisfied with their loan amount. It may be interesting to consider the fact that out of 371 samples, only 198 members (samples) are taking loan from PACB. At the time of primary data collection some of the samples, which have not taken any loans have reported that they are in need of a loan. From this one can easily conclude that the co-operative credit has not grown to the needs of farmers.

Table 7.3
Particulars of Loan Takers and their Loan Amounts

Sl. No.	*Loan Amount*	*Loan takers* *Satisfied*	*Not satisfied*	*Total*	*Chi-square value*	*P value*
	1	2	3	4		
1.	Below Rs. 5000	26 (36.6)	45 (63.4)	71 (100.00)	9.02	0.01098
2.	Rs. 5001- Rs. 10,000	31 (49.2)	32 (50.8)	63 (100.00)		
3.	Rs. 10,001 and above	40 (62.5)	24 (37.5)	64 (100.00)		
	Total	97 (49.00)	101 (51.00)	198 (100.00)		

Source: Primary data computed.
Note: Figures in brackets are percentage to row total.

Hypothesis-III: The Overdues are Inversely Related to Size of Land Holdings

For testing the third hypothesis, it has been reformulated to 'The overdues are equally distributed with respect to land size.

H_0 : The overdues are equally distributed with respect to land size.

Table 7.4 indicates that percentage of defaulters in each land holding group. Loan takers in the land holding group of upto 1 hectare and above made higher overdues in percentage wise (18.18%). It is 13.20 percentage and 12.90 percentage among 2-4 and 1-2 acres of land. The percentage of overdue respondents are very low within the high land holding groups than that of low land holding groups.

Table 7.4
Particulars of Borrowed Members and Overduers

Sl. No.	*Size of the land holding*	*Loan takers*	*Overduers*	*Percentage of column 3 to 2*
	1	2	3	4
1.	No land group	4	-	0
2.	Upto 1 hectare	44	8	18.18
3.	1-2	31	4	12.90
4.	2-4	53	7	13.20
5.	4-8	42	4	9.52
6.	8 and above	24	2	8.33
	Total	198	25	

Source: Primary data computed.

Table 7.5
Size of Land Holding and Overduers

Sl. No.	*Classification of land holding*	*Cases observed (O)*	*Expected (E)*	*Chi-square*	*P value*
	1	2	3		
1.	< 1 Hectare	8 (32.00)	5.00	4.800	0.3084
2.	1-2 Hectare	4 (16.00)	5.00		
3.	2-4 Hectare	7 (28.00)	5.00		
4.	4-8 Hectare	4 (16.00)	5.00		
5.	> 8 Hectare	2 (8.00)	5.00		
	Total	25 (100.00)			

Source: Primary data computed.
Note: Figures in brackets are percentage.

Calculated chi-square value is 4.80 with 4 degree of freedom. One can see that the significance level (p-value) is reported to be 0.3084. This high p-value leads us to accept the null hypothesis. Consequently one has to reach the conclusions that there is no direct relationship between the size of the land and overdues.

Hypothesis-IV: The Co-operative Movement has made Little in Mobilising Rural Savings

For testing the above hypothesis, it has been reformulated to the following form

H_0 : Savings in rural areas are equally distributed among various institutions.

The calculated chi-square value is 32.55 and p-value is 0.000, which is significant at 1% level, and therefore, one can reject the null-hypothesis.

Savings in rural areas are not equally distributed among the various institutions. It is seen in the table that among those households which have savings, 62 (44.93 per cent) samples selected LIC as their saving institution. But only 34 respondents (24.64 per cent) in the sample keep their savings in PACBs. Considering the fact that all our samples are members of the PACB, it can be easily concluded that PACBs are not performing well as institutions for mobilising savings. Out of 371 sample only 138 respondents are having savings and among this only 34 (9.17 per cent) of the total sample have chosen PACBs as their savings institution.

Table 7.6
Name of the Institutions and Savers

Sl. No.	*Name of the Institution*	*Cases observed (O)*	*Expected (E)*	*Chi-square*	*P value*
	1	2	3		
1.	PACBs	34 (24.64)	34.50	32.55	0.000**
2.	Commercial Banks	22 (15.94)	34.50		
3.	LIC	62 (44.93)	34.50		
4.	Others*	20 (14.49)	34.50		
	Total	138			

Source: Primary data computed.

Note: Figures in brackets are percentages.
* Others refers to relatives, friends, etc.

Hypothesis-V: The Co-operatives have Diversified the Rural Employment Activities

Particulars of loan issued for agriculture and allied activities by TPACB & KPACB as given below:

The above table gives a clear picture of the growth rate in loan amount under different heads. One can observe that growth rate in loan amount issued for allied agriculture is higher than (1.30) that of agriculture (1.20). The loans for allied agriculture certainly have the capacity to diversify the rural employment opportunity.

Table 7.7
Particulars of Loan Issued for Agriculture and Allied Activities

Sl. No.	*Purposes*	*TPACB & KPACB*			
		R-Square value	*F-value*	b_0	b_1 *(Growth rate)*
1		2	3	4	5
a.	Agriculture	.769	26.68**	741063	1.2049
b.	Allied - activities	.532	9.09*	49151.3	1.2893

Sources: Secondary data computed.
Note: * Values are significant at 5% level.
** Values are significant at 1% level.

The agricultural credit or seasonal credit constitutes working capital for the purchase of seeds, manure, other raw materials and a reasonable amount for the maintenance needs for the farmer and his family. This is also known as short-term (ST) loans, and constitute the core of production credit provided to the farmers. Bulk of the short-term loans are for agricultural purposes such as seasonal agricultural operations, i.e., for paddy, sugarcane, groundnut, ragi, bazra, jowar and maize etc. On the other hand, medium-term loans or loans for allied activities are similar. This type of loans are given for a period ranging from over 12 months to 5 years for purposes such as reclamation of land, purchase of live-stock, machinery and other implements, sinking of wells, purchase of cattle, bullock-carts, diesel engines and pump-sets for making minor improvements on land, and for sericulture, milch animals, calf rearing and bio-gas etc. In the analysis, data on loans issued only for allied activities have been used.

The above analysis has treated the sample as a whole. All the above mentioned hypotheses are tested for different societies to trace out the differences in performance in the following pages.

PART - II (SOCIETY WISE)

Hypothesis-I: Size of Credit is Directly Related to Size of Farm

The hypothesis has been reformulated as follows.

H_0: The size of credit do not differ according to size of farm.

The calculated chi-square value is 15.26, which is not significant (since the p-value is > 0.05). One can accept the null-hypothesis.

The calculated chi-square value is 49.73 and P-value is 0.000, which is significant at 1% level, one can reject the null-hypothesis.

Table 7.8
Size of Land and Size of Credit (Thamaraipakkam PACB)

Sl. No.	*Loan Amount →* *↓ Land Group*	*No loan*	*Below Rs.5000*	*Rs.5001-Rs.10,000*	*Rs.10,001- and above*	*Total*	*Chi-square value*	*P value*
	1	2	3	4	5	6		
1.	< 1 Hectare	11	12	11	10	44	15.26	0.227
		(25.0)	(27.3)	(25.0)	(22.7)	(100.0)		
2.	1-2 Hectare	5	4	4	7	20		
		(25.0)	(20.0)	(20.0)	(35.0)	(100.0)		
3.	2-4 Hectare	6	9	9	17	41		
		(14.6)	(22.0)	(22.0)	(41.5)	(100.0)		
4.	4-8 Hectare	5	6	8	12	31		
		(16.1)	(19.4)	(25.8)	(38.2)	(100.0)		
5.	8 and above	0	1	1	9	11		
			(9.1)	(9.1)	(81.8)	(100.0)		
	Total	27	32	33	55	147		
		(18.4)	(21.8)	(22.4)	(37.4)	(100.0)		

Sources: Primary data computed.
Note : Figures in brackets are percentage to row total.

Table 7.9
Size of Land and Size of Credit (Kalathure PACB)

Sl. No.	*Loan Amount →* *↓ Land Group*	*No loan*	*Below Rs.5000*	*Rs.5001-Rs.10,000*	*Rs.10,001- and above*	*Total*	*Chi-square value*	*P value*
	1	2	3	4	5	6		
1.	< 1 Hectare	48	9	1	1	59	49.73	.000**
		(81.4)	(15.3)	(1.7)	(1.7)	(100.0)		
2.	1-2 Hectare	24	11	5	0	40		
		(60.00)	(27.5)	(12.5)		(100.0)		
3.	2-4 Hectare	25	8	9	1	43		
		(58.1)	(18.6)	(20.9)	(2.3)	(100.0)		
4.	4-8 Hectare	10	6	8	2	26		
		(38.5)	(23.1)	(30.8	(7.6)	(100.0)		
5.	8 and above	5	3	5	5	18		
		(27.8)	(16.7)	(27.8)	(27.8)	(100.0)		
	Total	112	37	28	9	186		
		(60.2)	(19.9)	(15.1)	(4.8)	(100.0)		

Sources: Primary data computed.
Note: Figures in brackets are percentage to row total.

The chi-square test for the composite sample (Thamaraipakkam and Kalathure) has shown that in Table 7.2 there is a relationship between size of credit and size of farm. The same result has been repeated in the case of Kalathure society also. Interestingly there is no relationship between size of credit and size of farm in the case of Thamaraipakkam Society. One can easily conclude that in Thamaraipakkam society, the loans are not targeted towards land owners. Thamaraipakkam Society is providing different varieties of loans in which even landless people get opportunity to receive the loan. The Table No.6.15 has shown that in Thamaraipakkam Society 29 samples are receiving loans for consumer items. At the same time not even a single sample has received consumer loans in Kalathure Society. The same table reflected that 20 samples from Thamaraipakkam Society had procured loans for cattle but only two samples from Kalathure Society. For getting both these loans (Consumer and Cattle) land is not a criteria. Because of this, the relationship between size of credit and size of land is missing here.

Hypothesis-II: The Co-operative Credit has Not Grown to the Needs of Farmers

Ho: There is no relationship between satisfaction of loan takers and their loan amount in Thamaraipakkam PACB.

Table 7.10
Particulars of Loan Takers and their Loan Amount
(Thamaraipakkam PACB)

Sl. No.	*Loan Amount*	*Loan takers are Satisfied*	*Dissatisfied*	*Total*	*Chi-square value*	*P value*
	1	2	3	4		
1.	Below Rs.5000	19 (57.6)	14 (42.4)	33 (100.0)	1.12	0.57
2.	Rs.5001 -10,000	24 (68.6)	11 (31.4)	35 (100.0)		
3.	Rs.10001 and above	37 (67.3)	18 (32.7)	55 (100.0)		
	Total	80 (65.00)	43 (35.00)	123 (100.00)		

Source: Primary data computed.
Note: Figures in brackets are percentage to row total.

That calculated chi-square value is 1.12 and P-value is 0.57, one can accept the null - hypothesis.

The analysis given here shows that the calculated chi-square value is 1.06. As the P-value is 0.59, which is not significant at 5% level, the null hypothesis can be accept.

Table 7.11
Particulars of Loan Takers and their Loan Amounts (Kalathure PACB)

Sl. No.	*Loan Amount*	*Loan takers are* Satisfied	Dissatisfied	*Total*	*Chi-square value*	*P value*
	1	2	3	4		
1.	Below Rs.5000	7 (18.4)	31 (81.6)	38 (100.0)	1.06	0.59
2.	Rs.5001 -10,000	7 (25.0)	21 (75.0)	28 (100.0)		
3.	Rs.10001 and above	3 (33.3)	6 (66.7)	9 (100.0)		
	Total	17 (22.7)	58 (77.3)	75 (100.0)		

Source: Primary data computed
Note: Figures in brackets are percentage to row total.

Hypothesis-III: The Overdues are Inversely Related to Size of Land Holdings

Ho: The overdues are equally distributed with respect to land holdings.

Table 7.12
Size of Land Holding and Overduers (Thamaraipakkam PACB)

Sl. No.	*Classification of land holding*	*Number of overdue Persons*	*Expected*	*Chi-square value*	*P value*
	1	2	3		
1.	< 1 Hectare	4 (50.00)	2.67	1.75	0.42
2.	1-2 Hectare	-	-		
3.	2-4 Hectare	3 (37.50)	2.67		
4.	4-8 Hectare	1 (12.50)	2.67		
5.	> 8 and above hectare	-	-		
	Total	8 (100.00)			

Source: Primary data computed.
Note: Figures in brackets are percentage.

Calculate chi-square value is 1.75 and the P value is 0.42. One can accept the null-hypothesis. Consequently one can conclude that there is no relationship between overdues and size of land holdings.

Table 7.13
Size of Land Holding and Overduers (Kalathure PACB)

Sl. No.	*Classification of land holding*	*Number of overdue Persons*	*Expected*	*Chi-square value*	*P value*
	1	2	3		
1.	< 1 Hectare	4 (23.53)	3.40	0.94	0.92
2.	1-2 Hectare	4 (23.53)	3.40		
3.	2-4 Hectare	4 (23.53)	3.40		
4.	4-8 Hectare	3 (17.65)	3.40		
5.	> 8 and above hectare	2 (11.76)	3.40		
	Total	17 (100.00)	3.40		

Source: Primary data computed.

The analysis given here shows that the calculated chi-square value is 0.94 and P-value is 0.92, which is not significant at 5% level. The null-hypothesis can be accept. Consequently one can conclude that there is no relationship between overdues and size of land holdings. But, it is interesting to note that more number of persons are having overdues in Kalathure PACB.

Hypothesis-IV: The Co-operative Movement has made Little in Mobilising Rural Savings

Ho : Savings in rural area are equally deposited among various institutions.

Table 7.14
Name of the Institutions and Savers (Thamaraipakkam PACB)

Sl. No.	*Name of the Institutions*	*Number of Persons*	*Expected*	*Chi-square value*	*P-Value*
	1	2	3		
1.	PACB	29(36.71)	19.75	47.13	0.000
2.	Commercial Bank	5(6.33)	19.75		
3.	LIC	40(50.63)	19.75		
4.	Others	5(6.33)	19.75		
	Total	79(100.00)			

Source: Primary data computed
Note: Figures in brackets are percentages.

Since the calculated chi-square value is 47.13, which is significant at 1% level, one can reject the null-hypothesis. Savings made by samples from Thamaraikpakkam area are not equally distributed among the various institutions. Out of 79 samples, who have savings only 29 (36.71) samples of savings in PACB.

Table 7.15
Name of the Institutions and Savers (Kalathure PACB)

Sl. No.	*Name of the Institutions*	*Number of Persons*	*Expected*	*Chi-square value*	*P-Value*
	1	2	3		
1.	PACB	5(8.47)	14.75	10.35	0.0158
2.	Commercial Bank	17(28.81)	14.75		
3.	LIC	22(37.29)	14.75		
4.	Others	15(25.43)	14.75		
	Total	59(100.00)			

Source: Primary data computed.
Note: Figures in brackets are percentage.

Since the calculated chi-square value is 10.35 and the P value is 0.015 on can reject the null-hypothesis. Savings made by samples from Kalathure Society are not equally distributed among the various institutions. Out of 59 samples, who have savings just 5 (8.47%) samples of savings in Kalathure PACB.

From both the tables, one can derive the conclusion that rural people are not considering PACBs for depositing their savings. In Thamaraipakkam Society 36.71 percentage of the samples, who have savings selected PACBs for depositing their saving whereas only 8.47 percentage used PACBs for depositing their savings in Kalathure PACB. From the above analysis it can be derived that compared to Kalathure Society, Thamaraipakkam society is a better performing one in mobilising the rural savings.

Hypothesis-V: The Co-operatives have Diversified the Rural Employment Activities through their Loans.

Table 7.16 gives a clear picture of the growth rate in loan amount under different heads. One can observe from the table the growth rate in loan amount issued for agriculture, allied activities and non-agriculture. PACBs gives loan for agriculture purpose for purchase of fertilisers, pesticides, seeds, agricultural implements

Table 7.16
Particulars of Loans Issued for Agriculture, Allied Activities and Non-Agriculture by TPACB & KPACB

Sl.No.	Purposes	TPACB				KPACB			
		R-square value	F-value	b0 (Growth rate)	b1 square value	R-value	F-	b0	b1 (Growth rate)
	1	2	3	4	5	6	7	8	9
a	Agriculture	0.463	6.89*	534477	1.229	0.022	0.11	35761.8	0.9679
b	Allied Activities	0.603	12.16**	37150.2	1.3326	0.323	0.95	53852.2	0.8791
c	Non-Agriculture	0.873	55.17*	162297	1.4122	0.025	0.05	52866.5	1.1276

Source: Secondary data computed
Note: *Values are significant at 5% level
** values are significant at 1% level.

and other purposes. Allied-activities, which is related with purchase of bullocks and carts, camel and carts, Poultry farming, milch cattle, sheep rearing goat rearing, pig breeding and other purposes. Non-agriculture, which is meant for industrial purposes, jewel loans, consumption loans and other non-specified social purposes. Loans issued under the head non-agriculture is high in Thamaraipakkam society (1.412) than in Kalathure (1.127). Another interesting fact to be considered is that growth in loan amount issued for non-agriculture is higher than both agriculture and allied activities. And in Thamaraipakkam allied agriculture has a growth rate of 1.333 which is more than the growth rate of agricultural loans. This is clearly illustrates the fifth hypothesis that co-operatives have initiated steps to diversify the rural employment activities.

PART - III

Secondary data collected for various years are treated statistically in this section. Following portion of this chapter is an attempt to make a comparative study of the performance of the both societies in different sectors.

Table 7.17 discusses the growth of membership in Thamaraipakkam PACB and Kalathure PACB. One can observe that growth rate in membership of Kalathure Society is slightly higher than the Thamaraipakkam Society. But category-wise it differs. The category-wise such as below 1 hectare, 1-2 hectare and others, the growth rate is high in Thamaraipakkam than the Kalathure. In all other categories the growth rate shows an opposite picture.

Growth rate in loan amount is given in table 7.18. In Thamaraipakkam loan amount has a growth rate of 1.28 which is higher than the 0.976 growth rate of Kalathure. The same trend has shown in loan categories of short-term, medium-term and other loans. Different in growth rate with respect to medium-term loans are higher than the other category loans. Medium-term loans are mainly for allied agriculture and the same picture can be seen in Table 7.7 (Hypothesis -V), where loan amount for allied agriculture are is much high in Thamaraipakkam than in Kalathure.

When these two tables are looked into deeply, one is surprised to find a misleading trend in the growth of membership. The growth rate in membership seems to be higher in Kalathure where

Table 7.17
Growth of Membership Both in TPACB & KPACB

Sl. No.	Variable	TPACB				KPACB			
		R-square value	F-value	b0	b1 (Growth rate)	R-square value	F-value	b0	b1 (Growth rate)
	1	2	3	4	5	6	7	8	9
1.	Below 1 Hectare	0.524	8.82*	144.385	1.1064	0.979	236.12**	558.128	1.0161
2.	1-2	0.499	7.96*	266.792	1.0441	0.996	1200.92**	339.357	1.0203
3.	2-4	0.165	1.58	344.268	0.9604	0.836	25.5**	180.600	1.0159
4.	4-8	0.172	1.66	88.2630	0.9445	0.875	35.12	122.841	1.0351
5.	8 and above	0.625	13.34**	89.2219	0.8640	0.245	1.63**	26.1350	1.0614
6.	Agri. Labourer	0.001	-	133.536	0.9927	0.949	92.62**	175.153	1.0478
7.	Rural Artisan	0.160	0.59	26.0832	0.9691	0.844	27.09**	75.4667	1.0494
8.	LAL	0.160	1.34	29.8713	1.1862	0.962	127.30**	199.895	1.0511
9.	Total members	0.084	0.74	1141.67	1.0207	0.973	181.22**	1663.18	1.0295

Source: Computed Secondary data by TPACB & KPACB.
Note: *Values are significant at 5% level.
** Values are significant at 1% level.

Table 7.18
Loans Issued by TPACB & KPACB

Sl. No.	*Variable*	*TPACB*				*KPACB*			
		R-square value	*F-value*	*b0*	*b1 (Growth rate)*	*R-square value*	*F-value*	*b0*	*b1 (Growth rate)*
	1	2	3	4	5	6	7	8	9
1.	Short-term loan	0.463	6.89*	53447.7	1.2291	0.022	0.11	357618.0	0.9679
2.	Medium - term loan	0.603	12.16*	37150.2	1.3326	0.323	0.95	53852.2	0.8791
3.	Others	0.873	55.17**	162297	1.4122	0.025	0.05	52866.5	1.1276
4.	Total amount	0.888	63.30**	820350	1.2824	0.013	0.07	481726.0	0.9760

Source: Computed Secondary data by TPACB & KPACB.
Note: *Values are significant at 5% level.
** Values are significant at 1% level.

Table 7.19
Particulars of Issued Loan Amount, Collection & Balance by TPACB & KPACB

Sl. No.	*Variable*	*TPACB*				*KPACB*			
		R-square value	*F-value*	*b0*	*b1 (Growth rate)*	*R-square value*	*F-value*	*b0*	*b1 (Growth rate)*
1		2	3	4	5	6	7	8	9
1.	Loan issued	0.967	237.37**	830009	1.3018	0.089	0.49	370481	1.0805
2.	Collection	0.412	5.61*	696882	1.2300	0.477	4.56	229030	1.1548
3.	Outstanding	0.980	400.74**	1016846	1.2637	0.785	18.25**	810631	1.1061

Source: Computed Secondary data by TPACB & KPACB.
Note: *Values are significant at 5% level.
** Values are significant at 1% level.

Table 7.20
Particulars of Savings by TPACB & KPACB

Sl. No.	Variable	TPACB				KPACB			
		R-square value	F-value	b0	b1 (Growth rate)	R-square value	F-value	b0	b1 (Growth rate)
	1	2	3	4	5	6	7	8	9
1.	Fixed deposit	0.911	81.46	9586.88	1.9169	-	-	-	-
2.	SB A/c	0.015	0.12	50356.5	1.0544	-	-	-	-
3.	RD A/c	0.540	9.38	285.293	1.5944	-	-	-	-

Source: Computed Secondary data by TPACB & KPACB
Note: *Values are significant at 5% level.
** Values are significant at 1% level.

Table 7.21
Total Membership, Actual Loan Users and OD Persons

Sl. No.	Variable	TPACB				KPACB			
		R-square value	F-value	b0	b1 (Growth rate)	R-square value	F-value	b0	b1 (Growth rate)
	1	2	3	4	5	6	7	8	9
1.	Total membership	0.086	0.75	1136.61	1.0211	0.953	101.79**	1662.74	1.0301
2.	Actual loan user	0.685	17.37**	236.210	1.1964	0.011	0.06	81.4224	1.0243
4.	Number of OD person	0.010	0.08	153.455	0.9894	0.013	0.06	215.112	1.0098

Source: Computed Secondary data by TPACB & KPACB.
Note: ** Values are significant at 1% level.

as the loans have not grown at a faster rate. The membership might have been boosted to capture the elected posts of the cooperative society and hence one can infer that there seems to be more political intervention at Kalathure than at Thamaraipakkam.

Table 7.19 gives a clear picture of amount of loan issued, amount collected and amount of overdues. The growth rate in all this, are higher in Thamaraipakkam than in Kalathure. It can be concluded that the reason for a higher growth rate of outstanding in Thamaraipakkam is mainly because of the higher growth rate in the amount of loan issue. The outstanding balance does not mean over dues.

Table 7.20 gives a clear picture about the deposit status of the members in Thamaraipakkam PACB and Kalathure PACB. But in this table, data are available only in Thamaraipakkam PACB. It has been shown that recurring deposit (1.59), which is higher than the saving account (1.05) and fixed deposit (1.92) in Thamaraipakkam PACB only.

Growth of total membership, actual loan users and number of OD persons are given in the above table 7.21. In the case of total membership Kalathure society has achieved a growth rate of 1.03, which is slightly higher than the growth rate of Thamaraipakkam Society (1.02). But the growth rate of actual loan user is higher in Thamaraipakkam (1.19) than in Kalathure Society (1.02). Considering the number of over due persons one can observe that the growth rate in overdued persons is higher in Kalathure Society (1.00) than in Thamaraipakkam Society (0.989).

From all these analyses one can infer that if a PACB is to perform well, it should have little political intervention and effective borrowers should be more in number. The members should also repay the loans in time so that the PACBs can function well. Thus ultimately the efficiency of the PACB rests on the cooperative behaviour of the members.

Chapter - 8

SUMMARY AND CONCLUSION

The Co-operative banking in India has grown in size and volume. When the movement was started in 1904 only agricultural co-operatives were popular. At present, co-operative banks are organised in many sectors. A special feature of co-operative banks in India is that they have a federal structure, the units ranging from primary level to national level. Co-operatives as a form of organisation and institutional strategy are best suited not only to alleviate poverty but also to ensure growth with social justice. The credit co-operatives represent a nation-wide network over the entire rural India and there is no other organisation which is so well spread in the country.

The Co-operatives have played a very important role in the country's agriculture system. Credit co-operatives have a considerable share in the success of Green Revolution. Similarly, in bringing about the white Revolution, co-operative's role is well known to the world. To imagine Indian agriculture without co-operatives is really difficult.

The essential ingredient of agriculture development is availability of adequate credit. The adoption of new technologies and modern inputs have increased the demand for agricultural credit which is considered to be the life blood of agricultural development.

Agricultural credit in India is characterised by dual structures i.e., institutional as well as non-institutional. The institutional credit mainly comprises of co-operatives, commercial banks, regional rural banks and government, while the non-institutional credit consists of moneylenders, landlords, commission agents, traders, relatives and friends.

Institutional finance plays a vital role in the development of agricultural and allied activities in India. Among the institutional agencies providing credit to agricultural sector, co-operatives play a pivotal role. They are viewed as the most suitable and viable agencies for rural development under indian conditions. The co-operative credit has a three-tier structure, namely, the state co-operative banks at the Apex, the district central co-operative banks (DCCBs) at the middle or district level and the primary agricultural co-operative banks (PACBs) at the lower or village level.

The district central co-operative banks play a vital role in the disbursement of co-operative credit. The progress of co-operative credit programmes largely depends on the financial strength and viability of these banks. But a good majority of DCCBs are in financial distress, because of operational inefficiency. The co-operative as tool of rural development have not fully succeeded in accomplishing the objectives for which they had been established.

The primary co-operative banks and the co-operative movement itself are at cross roads. The economic reforms and the subsequent financial reforms may expose the weakness of co-operative banks. They may have to forego subsidised credit, may have to face mounting overdues, may have to meet the widening credit gap between demand for agricultural credit and supply through co-operative banks and may have to succeed in their failure to mobilise rural savings. The success of the co-operative agricultural credit depends upon the success of primary co-operative banks and hence the study is to review the working of two co-operative societies identified on the advice of Vellore district co-operative authorities and substantiated by secondary data. It has been decided to select one 'bad' and one 'good' PACB for the present study and sample respondents also have been selected adopting proportionate random sampling method. Ten percent of the members of the banks were selected as sample households.

CO-OPERATIVE MOVEMENT-A REVIEW

The history of modern civilization is, in fact, the history of co-operation, for without it, social and economic progress would not have been possible. The genesis of modern co-operatives can be traced back to the first half of the 18th century. Robert Owen, a poor welsh man, the "father of modern co-operative movement", was the first to propound and propagate the ideals of co-operation in 1793. From England, the movement gained momentum and spread to all the countries round the globe, whether developed or developing. As G.D.H. Cole observed: "Today co-operation is a world-wide movement".

Co-operation is not in any way new to India. It has a fairly long history in India. It has been in existence since times immemorial. However, the passage of the Co-operative Societies Act in 1904 was the first milestone in the movement. The Act of 1919 gave great stimulus to the movement. It made some progress during pre-Independent era. Co-operation has been assigned a notable role in bringing socio-economic transformation though the process of democratic planning with an accent on assisting the weaker sections in rural areas. A salient feature of the Indian co-operative movement is that it has not only numerically expanded but also continuously diversified its activities over the years, leaving no sector of economy untouched. But, despite its ambitious plans and diverse developments, the co-operative sector has faced numerous problems which have checked its optimum growth. The co-operative movement in India continues to be a state promoted, controlled and guided movement and not a people's voluntary movement. Though 'co-operation' is a state subject under the constitution of India, the central government guides, assists and co-ordinates the movement in the states.

The All-India Rural Credit Survey Committee observed that, during 1951-52, there were only two important sources of rural Credit, namely, the moneylenders and the borrower's relatives, who between themselves accounted for about 84 percent of the total credit. The government and the co-operatives accounted for 6.4 percent and commercial bank for just 0.9 percent.

The Rural Credit Co-operative Institutions (RCCIs), one of the strong arms of Rural Financial Institutions (RFIs), have made significant strides in delivering credit for agriculture and rural development. Beginning with the first decade of the current century, the co-operative movement has taken deep-roots and culminated in to a massive rural co-operative institutional structure. Today, co-operatives form the major component of the rural credit system in India. Of the total credit (including short-term, medium-term and long-term) disbursed during the year 2001-2002, the share of co-operatives was 40.56 percent, i.e., about half of the total credit disbursed. The share of the commercial banks and regional rural banks, the other two institutional players in the rural credit market, together finance 59.44 percent of institutional rural credit. In other words, co-operatives still constitute a single dominant segment of the rural credit market.

Several studies were made by individual researchers, institutions and government agencies at the macro, regional, national and international levels, to evaluate the working of co-operatives from time to time. The Government of India and Reserve Bank of India, were to play an active role in the supply of rural credit. Since the launching of First Five-Year Plan in India in 1951, many expert committees were appointed from time to time to suggest methods to improve the flow of agricultural credit from institutional sources. A few of them are: All-India Rural Credit Survey Committee (1951), Mehta Committee (1959), Patel Committee (1961), Mirdha Committee (1964), All-India Rural Credit Review Committee (1966), Santhanam Committee (1969), Venkatappiah Committee (1969), Bawa Committee (1971), All-India Debt and Investment Survey (1971-72), Madhava Das Committee (1973), Hazari Committee (1975), Report of the National Commission on Agriculture (1976), Kamat working group (1976), Sivaraman Committee (1979), Khusro Committee (1989), Narasimham Committee (1991), Jagadish Capoor Committee (1999), Report on currency and Finance (1998), and Report on Trend and progress of Banking in India (1998-99). All these reports have pointed out both the strength and weakness of co-operative credit institutions. A review of the studies made by individual researchers and various committees have revealed that co-operatives have succeeded partially in fulfilling their objectives and are currently facing various problems.

Vellore district has been chosen for the present study and the district has experienced an appreciable change in agricultural development. Decadal growth rate of population of the district has been reduced from 15.14 in the period between 1981-1991 to 15.09 for the period 1991-2001. At the same time decadal growth rate of the state as a whole has been reduced from 15.39 in 1981-1991 to 11.19 in 1991-2001. Thus the decadal growth rate of the district is much higher compared to state as a whole. Vellore district is predominantly agriculture oriented and a greater proportion of the economically active population is involved in agriculture and allied activities. In 1961 census 51.44 percent of the population were cultivators. It has come down to 25.48 percent of the population by 2001 census. It has been reduced by 28.04 percent point over four decades. During 1999-2000, 80 percent of the land holdings are of size less than 1 hectare, and 14 percent are of size between 1 and 2 hectares. While 5 percent of the land holdings fall in the category 2 to 5 hectares, only 1 percent of the holdings are of size above 5 hectares. One can easily say that this district has more number of marginal holdings which cannot assure adequate income to the family.

Agricultural activity in the district is entirely dependent on the seasonal rainfall. Data indicate that much has to be done to increase the productivity of sugarcane, pulses and cotton. A significant effort has to be taken in sugarcane and pulses. The data pertaining to the particulars of financial institutions in vellore district by the end of March 2002, reveal that there are 29 banks working in the district and among them 18 are nationalised banks, 11 private banks and 1 DCCB with 35 branches and 185 PACBs in the district.

Co-operatives form the major component of the rural credit system in Tamil Nadu as well India. Of the total credit (including short-term, Medium-term, long-term) disbursed during the year 1989-90, the share of co-operatives was 50.87 percent, i.e., about half of the total credit disbursed. But it has shown a declining trend in 2001-2002 i.e., 40.56 percent. The deposits of the TNSC bank increased from Rs.387.24 crores in 1989-90 to Rs.2676.11 crores in 1999-2000. The fresh loans advanced by TNSC banks reached the peak of (short-term only) Rs.62020. 38 lakhs in 1998-99 from mere Rs.49121.00 lakhs in 1997-98.

In July 1904, the first society to be registered was the Tirur rural society in Chengalpattu district of Tamil Nadu. The deposits of the VDCCB in the year 1917-1918 was Rs.3,10,938 and in the year 2001-2002, the deposits rose to 275.73 crores. The loan outstanding was only Rs.3,65,524 in the year 1917-1918 and it has increased to Rs.321.00 crores in the year 2001-2002.

The role of co-operative banks as well as commercial banks for the distribution of agricultural credit has steadily grown up in the vellore district. During 1988-89, the share of crop loan through commercial banks was 26.00 percent and it rose to 57.65 percent by 2001-2002 in the vellore district. The co-operatives which accounted for 74.00 percent in 1988-89 declined to 42.35 percent for 2001-2002 in vellore district. In the year 1994-95, there been neither profit, nor loss. In the year 2000-2001 VDCCB incurred a loss of Rs.369.69 crores. The profit earned is Rs.4.00 lakhs during the year 2001-2002. In the year 1994-95 there were 423346 member in VDCCB and it has increased by 109649 members in 2001-2002 and at the same time actual loan users increased from 126211 members in the year 1994-1995 to 213071 members in the year 2001-2002.

In the year 1990-91 the loan amount issued by all PACBs in Vellore district amounted to Rs.2802.17 lakhs and it has increased to Rs.19717 lakhs by the year 2001-2002. In the year 1990-91 short-term loan accounted for 53.74 percentage of the total loan issued. Percentage of the short-term loan in the total loan amount has declined to 47.57 percent by the year 2001-2002. Overdues in VDCCB, is continuously on downward trend since the bank has taken strict measures for reducing the percentage of overdues. It shows that overdues percentage to credit was 52.03 percent in 1990-91 which decreased to 34.00 percent in 2001-2002. It is clear that deposits of all PACB membership at VDCCB, has grown up considerably over the 12 years period. The deposit was Rs.529.84 lakhs in the year 1990-91 and has increased by Rs.11686.23 lakhs to reach Rs.12216.07 lakhs by the year 2001-2002. Having seen the success and failure of Vellore Central Co-operative Bank, the working of sample PACBs is discussed.

The census data reveal that villages of Thamaraipakkam PACB area have experienced less population growth than the Kalathure PACB villages. One can observe that larger geographical area is covered by Thamaraipakkam PACB than Kalathure PACB. There is no higher Secondary school in these villages. There is only one local general library at Thamaraipakkam. During the period 2001-2002, paddy and sugarcane were the most important crops in terms of area under cultivation in these both PACB villages.

RESEARCH FINDINGS

The members of TPACB who belongs to below 1 hectare steadily increased from 15.08 percent in 1989-90 to 27.49 percent in 2001-2002. This confirms that higher percentage coverage of borrowing members were from the land holders belonged to below 1 hectare. This can be taken as an indicator for good performance of Thamaraipakkam PACB, which helped the poor and needy farmers. The total number of membership by the landholders below 4 hectare is higher than that of the land holders having 4 hectare and above. Kalathure which is another PACB taken for the present study shows that those who owned below one hectare formed 32.90 percent in 1989-90 and this has come down to 29.87 by 2001-2002. Loan amount for non-agricultural purposes constituted 28.15 percent in the year 1989-1990 and has increased to 35.76 percent in 2001-2002 in the case of Thamaraipakkam PACB. The same has increased from 13.34 percent in 1992-93 to 22.47 percent in 1996-97 of Kalathure PACB and has reached the lowest level of 0.52 percent in 2001-2002. The total amount of loan issued by TPACB has increased from Rs.1306482 to 13237935 between 1989-96 to 2001-2002 and that of KPACB from Rs.472144 to 2010825 for the same period.

PACBs receive their deposits from members and non-members. Fixed deposit is an important source of deposit for the co-operative banks. In the case of fixed deposit Thamaraipakkam PACB has showed an appreciable upward trend from 1989-90 to 2001-2002. However the PACBs have not mobilised the rural savings available in the villages and the savings have been deposited with other institutions. There has been marginal change in the membership of Thamaraipakkam PACB, from the base year 1989-90 to 2001-2002. There are upward trend in the actual loan user and number of persons having overdues between 1989-90 and 2001-2002 in Thamaraipakkam PACB. The total members have increased in Kalathure by 40.39 percent. The index number has increased to 140.39 in terminal year 2001-2002. If 1988-89 is taken as base year effective loan users had declined in the period between 1989-1990 and 1997-1999.

The samples have been analysed to test the hypotheses and the test has established that there exists a positive relationship between size of land holdings and the loans issued. It has been found that 65.95 percent of the landless, small and tiny land owners who own less than 1 hectare have not received any loan. However, only 23.25 percent of those who own more than 4 hectares have not received loan.

Again it was found through the analysis that the co-operative credit has not grown to the needs of farmers. As a matter of fact out of 198 loan takers, 101 have expressed dissatisfaction over the co-operative credit and the dissatisfaction is highly significant as the P value is 0.01. Further as the loan amount increases the percentage of dissatisfied persons decreases.

It has been established through primary data that there exists a gap between demand for credit and supply of credit through co-operative banks. Further, over dues are threatening the very existence of PACBs. However, it was found that there is no such direct relationship between size of landholdings and the overdues. The PACBs have not succeeded in mobilising rural savings and the failure is highly significant as the P value is 0.000.

The analysis of two PACBs also establish the same facts. The same hypotheses have been tested at the level of PACBs using both secondary and primary data. The first hypothesis, that the size of credit do not differ according to size of farm is accepted for Thamaraipakkam, but rejected for Kalathure PACB. Secondly, the hypothesis that there is no relationship between satisfaction of loan takers and their loan amount is accepted at both the PACBs. The third hypothesis that overdues are equally distributed with respect to size of land holdings is also rejected in both the PACBs. The fourth one viz. Savings in rural area are equally deposited among various institutions is rejected in case of both the PACBs and the lost hypothesis that the co-operatives have diversified the rural employment through their loan is accepted in case of both PACBs.

The 'good' and 'bad' PACBs reveal their distinctive characteristics in the analysis. Thamaraipakkam PACB which is described as 'good' PACB has less non-effective members, more mobilisation of rural savings, higher amount of loan, larger proportion of loan for agricultural allied activities and non-agricultural purposes and in the total volume of credit extended.

The working of the two co-operative banks show distinctly different characteristics on many respects. Though majority of the members of both PACBs belong to the category of small and marginal farmers, those own between 1 and 4 hectares are larger at Thamaraipakkam PACB compared to Kalathure PACB. Landless are more among the members of Kalathure PACB.

The income data for the members of both PACB show that more than 40.52 percent belong to the income group Rs.10,000 and above where as in Kalathure this group form only 28.44 percent. The low income profile of Kalathure members have to be taken into consideration before talking of repaying capacity and savings potentialities of the members.

The actual members have experienced a small growth rate (b1) of 1.0211 in TPACB where as the increase was much higher in Kalathure (b1 = 1.0311). However actual loan users have shown an upward trend in TPACB where as the actual loan users have declined in KPACB with exception during certain years.

In the same way the growth of loans issued in these two PACBs have also differed during the period of study. The short term loan has experienced a growth rate (b1) of 1.229 or a percentage of variation of 22.9 percent at TPACB, medium term loan issued has shown growth rate of (b1) 1.333 and other loans have experienced a growth rate (b1) of 1.412 where as the same b1 rates are 0.967, 0.879 (negative) and 1.127 respectively. Thus the growth of loan in percentage variation differs very widely between TPACB and KPACB.

Again, loan issued for non-agricultural purposes also show a visible difference between TPACB and KPACB. The loan for non-agricultural purpose has shown a growth rate of (b1) 1.412 during the study period in TPACB, where as in KPACB this is much lesser. The loan to the agricultural operations have indicated a growth rate of (b1) 1.229 in TPACB and 0.967 (negative) in Kalathure PACB.

In short if b1 (growth rate) is taken as indicator, loan issued has shown a growth rate of 1.301 (30.1 percentage variation)), collection has shown b1 growth rate of 1.230 (or 23.0 percentage variation) and outstanding loans also have shown b1 growth rate of 1.263 or 26.3 percent variation in case of TPACB during the period of analysis. If the same variables are considered for KPACB, the b1 values are 1.080, 1.154 and 1.106 respectively. In other words loan issued show a percentage variation of 8 percent, collection 15 percent and out standing loans 10 percent during the period of analysis.

If mobilisation of rural savings is taken as indicator, again TPACB has shown better in mobilising the rural savings. More dissatisfied persons are there in Kalathure PACB compared to Thamaraipakkam PACB. It has been revealed during survey that there has been more political intervention at Kalathure PACB compared to Thamaraipakkam PACB.

The members of KPACB are not willing to approach the PACB for loans as they feel many problems such as finding surety going for auction incase of failure to repay the loan, high penal rate of interest already having outstanding balance etc. Thus whatever indications suggested by co-operative officials are taken up, Kalathure PACB seems to have performed poor.

SUGGESTIONS

On the basis of the above findings, a few suggestions are given which may improve the performance of co-operative banks and the borrowers to derive maximum benefits.

Politicisation of administration is the hallmark of the co-operative credit system today. The political party members are able to extract more loans from the co-operative banks. It should be made as a people's organisation with least politicisation and bureaucratisation. KPACB is the standing example to prove the case that too much of politicisation may lead to the creation of a bad PACB.

There should be proper supervision over the working of the branches and the PACBs. To this extent, the supervisory staff should be made effective and be trained properly.

The administrative system should be made more professional and the credit management training should be given to them. The recruitment and training should be modified to give them a competitive skill to work in different environment.

Many of the respondents felt that the interest charged by the co-operatives and the penal rate charged are too high. This may be a reason for overdues. The three-tier system make the interest high according to respondents. The apex body borrows from NABARD at lesser rate of interest and lend to CCB at 6.25 percent. The CCB lends to PACB at 9 to 10 percent depending upon the size of loan and PACBs lend to members at rates of interest ranging from 12 to 14 percent depending upon the size of credit. It is felt that by abolishing the APEX body and NABARD lending directly to CCBs, the lending rates of PACBs can be brought down.

Official procedures of getting the loans should be streamlined. The existing system of requiring third person surety should be removed.

Deposits are important both for the customers and for the banks. Farmers do not get monthly income. Their income is only seasonal i.e., they have some surplus by sale of produce after each harvest season. There may be two (or) three seasons in an area. It is suggested that for farmers three-monthly or four-monthly recurring deposit schemes should be introduced. The period of these deposits should coincide with the major harvest seasons of those areas, which are served by these banks.

The crop insurance scheme should be widely published and made applicable to all part of the country. PACBs should come forward to pay a part of the premium of insurance policy, instead of government giving rebates in penal interest rates. This would help farmers to take interest in this. This would enable the farmers to protect themselves from the natural calamity like floods, loss in operation and other drought situation. It will reduce the repayment problems and avoid the overdue.

The loan amount should be on the basis of real cost of cultivation. In modern times, the cost of cultivation has increased manifold. The amount of loan given to members is quite inadequate so that members are attracted towards moneylenders for getting rest of the need fulfilled.

Accumulation of overdue constitutes a major syndrome affecting the working of district central co-operative banks as well as PACBs. The main reasons for increasing overdues are defective loaning policies, including untimely loan disbursement, under finance, misuse of loans by the borrowers, unforeseen natural calamities, such as droughts, floods, etc. This can be avoided if marketing of produce linked with loan as in case of co-operative sugar mills. Thus a strong and well established crop insurance system alongwith well knit cooperative lending systems with marketing systems as in case of cooperative sugar mills can help to overcome overdues.

It should be born in the minds of all those who are the members and officials of the co-operative banks that co-operatives should not be an end but it should be a means to improve the agricultural sector as well as quality of life of indian farmers. To this end, the Vellore District Central Co-operative Bank in general, and the Thamaraipakkam PACB and Kalathure PACB in particular, should work effectively to achieve the main objectives of developing rural India.

FURTHER RESEARCH

The study is limited to two PACBs and hence many more studies on the working of good and bad PACBs have to be made to arrive at a general conclusion for policy formulation.

A separate study can be made on politicalisation of cooperatives and its impact on a democratic institution like cooperatives.

A study can also be made on the opinions of rural people about their preference to nationalised commercial banks and not cooperatives to deposit their savings. This study will be more helpful to overcome the financial weakness of cooperatives.

A separate study can also be made on the credit gap experienced by farmers and the mechanism adopted by them to meet this gap. A study can also be made on recapitalisation to make PACBs viable and the impact of New Economic Policy where competitive financial system is to emerge on the working of cooperatives.

When the entire discussions are pulled together, certain general conclusions can be drawn. It can be said that the co-operative credit institutions are the best suited financial institutions in the present agrarian environment. The present agrarian structure in the study area is dominated by more number of unviable small and marginal farmers. They are affected by production uncertainty and price uncertainty. The cost of cultivation is also on increase and if much talked steps viz. reduction of subsidies if not removal of subsidies is to take place the cost of cultivation is likely to increase still further. India as a member of WTO has to fulfill certain obligations and hence the import of agricultural products may impose a restriction on Indian farmers. There are indications from research findings that the yield per hectare has shown a decline and this trend is more visible in Punjab. The cost of cultivation is on increase and farmers are not in a position to meet their debt obligations. Thus the agrarian structure seems to be on the cross - road. The domination of small and marginal farmers complicate the issue. The environment in which the co-operative credit banks have to function are more risky. Majority of them are poor marginal and small farmers who have to face both production and price uncertainties. They need a cheap credit to invest in a risky productive venture and in the absence of crop insurance, the risk of farmers are further heightened. Under these conditions cooperative credit institutions have to play a dominant role. However, the discussion points out that imposing an unequal land holding structure on a democratic institution (co-operatives) will only lead to unequal credit structure.

The problem of the cooperative credit institutions seems to be the problem of lack of cooperation from its own members and political intervention. The lack of cooperation from members leads to overdues and depositing the available savings with other agencies and not with PACBs. Thus the members should be made to realise that it is their bank and it is they who manage it and the effective functioning of cooperative institutions depend upon the prompt repayment of loans. The credit gaps must be met by cooperative institutions and otherwise it will drive the people to other agencies in a multi-credit agencies system.

If the co-operative credit movement is to succeed their own members have to cooperate and if the movement fails the agrarian economy will fail and the members may fail to improve their economic condition. The success rate of co-operative movement rises with the co-operation of members the members fall with the fall of PACBs. If the members are alert sincere and honest, they can even control the corrupt officials if any or self centered elected person who are causing damage to the movement. The PACB to be good or bad is not the fault of institutions but the fault of people who administer it and the fault of people who participate in its operation as members. Perhaps ultimately it is left to their members to make a PACB good or bad.

APPENDICES

APPENDIX 1

Number of Cooperative Institutions in Tamil Nadu

Sl. No. (1)	*Type* (2)	*1994-95* (3)	*1995-96* (4)	*1996-97* (5)	*1997-98* (6)	*1998-99* (7)	*1999-2000* (8)
1.	Primary Agricultural Credit Societies	4482	4495	4495	4520	4534	4537
2.	Central Cooperative Banks	22	22	23	23	23	23
3.	State Cooperative Apex Bank	1	1	1	1	1	1
4.	Primary Cooperative Land Development Banks	181	181	181	181	181	181
5.	State Cooperative Land Development Bank	1	1	1	1	1	1
6.	Cooperative Urban Bank	129	127	127	127	129	129
7.	Employees Cooperative Credit Societies	1755	1770	1793	1813	1826	1864
8.	Primary Cooperative Consumer Stores	3159	3619	3600	3595	3599	3557
9.	Consumer Cooperative Wholesale and District Supply and Market Societies and Super Markets	31	30	30	31	36	36
10.	State Conumer Cooperative Federation	1	1	1	1	1	1
11.	Primary Cooperative Marketing Societies	117	104	112	116	117	117
12.	State Cooperative Marketing Federation	1	1	1	1	1	1
13.	Cooperative Printing Works	21	21	20	23	26	26
14.	Washermen Cooperative Societies	8	4	3	3	3	3
15.	Hill Tribes Cooperative Marketing Societies	9	8	6	6	5	5
16.	Barbers Cooperative Societies	5	4	5	4	4	4
17.	Salt Workers Cooperative Societies	15	12	12	13	12	12
18.	Labour Contract Cooperative Societies	217	184	174	177	181	182
19.	Societies for Weaker Sections (LAMPS)	12	12	18	11	16	16

Source: Tamil Nadu—An Economic Appraisal, 2001-2002.

APPENDIX-II

Co-Operative Credit for Agricultural Sector: With Special Reference to Vellore District-Tamil Nadu

INTERVIEW SCHEDULED

Village	:	
Taluks	:	Arcot / Arakkonam
District	:	Vellore

1.0 Identification Particulars

1.1	Respondent's Name	:	
1.2	Category	:	LF/SF/MF/AL/LAL
1.3	Age	:	
1.4	Sex	:	Male/Female

2.0 Social Classification

2.1	Religion	:	
2.2	Respondent belongs to	:	SC:1, BC:3, FC:5 ST:2, MBC:4

2.3 Particulars of Household members

Sl. No	*Name*	*Relationship*	*Age*	*Sex*	*Marital Status*	*Educational Qualification*
1.						
2.						
3.						
4.						

3.0 Literacy Level of the beneficiary

(a) illiterate ☐

(b) able to read and write ☐

(c) Primary/middle Education ☐

(d) S.S.L.C. ☐

(e) HSC ☐

(f) Degree level ☐

(g) P.G and ☐

(h) Any professional ☐

4.0 Occupational status and their annual income

S. No	*Name of the Occupation*	*Number of Earning members* *Male*	*Female*	*Children*	*Total*	*Annual Income*
1.	Agriculture					
2.	Allied activities					
3.	Business					
4.	Industry/Company					
5.	Any employee with State/Central					
6.	Other Specify					
	Total					

5.0 Particulars of total land and cultivated area

Sl. No	*Particulars*	*Wet Land*		*Dry Land*	
		Total	*Under the cultivation only*	*Total*	*Under the cultivation only*
1.	Owned Land				
2.	Tennancy Lease in				
3.	Tennancy Lease out				
4.	Total				

6.0. Sources of irrigation

Sl. No	*Sources*	*Availability (Please ✓ tick)*	*Total*
1.	Canal		
2.	Tank		
3.	Well (a) Pumpset (b) Oil Engine		

7.0 Particulars of savings (Previous year) Deposits

Sl. No.	*Sources of savings*	*Amounts (in Rupees)*
1.	PACB	
2.	Commercial Bank	
3.	Relatives others	

8.0 Are you member of Co-Operative bank? : Yes / No

8.1 If yes, have you got any loan : Yes / No

8.2 If yes, Whether loan is in time : Yes / No

8.3 If no, please state specify? : 1

2

3

8.4 Purpose for which the loan is sought by PACB

Sl. No	*Particulars season (Amount)*	*Last Agricultural cumulative*	*Outstanding*
(a)	Purchase of fertilizers/manure		
(b)	Purchase of seeds		
(c)	Purchase of cattle		
(d)	Purchase of Agri-implements		
(e)	Others (Non-Agri)		

8.5 Are you satisfied with loan amount getting from PACB : Yes / No

8.6 If No, any alternative sources for which is agricultural business, Namely

Sl.No	*Alternative Sources*	*Agriculture*	*Non-Agriculture*
(a)	Commercial Bank		
(b)	RRB		
(c)	Moneylenders		
(d)	Relatives		
(c)	Friends		

9.0 If you have not used PACB for the last agricultural seasons-why?

1. own money 2. not cultivated 3. over due

4. private 5. others

10.0 Are you getting any subsidy from PACB about your agricultural business (Previous Year)

Sl.No	*Item of subsidy/ Grant*	*Total loan Amount Rs.*	*Subsidy Amount Rs.*
(a)	Seeds		
(b)	Fertilizers		
(c)	Pesticides/insecticides		
(d)	Agri-Implements		
(e)	Motor pump-set		
(f)	Cattle		
(g)	Others (specify)		

11.0 Why did you approach the PACB for financial assistance for agriculture business?

Sl. No	*Particulars*	*Please tick your options*
(a)	Low interest rate	
(b)	Easy terms of repayment	
(c)	Near to residence	
(d)	To avoid unnecessay risk	
(e)	To enjoy subsidy	
(f)	To avail consultacy and quidance service	
(g)	Other (specify)	

12.0 Did you come across any problem in getting the loan from PACB

Yes: ☐

No : ☐

12.1 If yes, what are the problems

Sl. No	*Types*	*Please tick which is suitable*
(a)	To deposit any amount	
(b)	Any Third person surity	
(c)	Any delay in sanctioning the loan	
(d)	Any political involvement	
(e)	Any communal problem	
(f)	Any commission	
(g)	Any others (specify)	

13.0 Repayment performance

13.1 Did you repay the loan amount regularly? Yes: ☐

No : ☐

13.2 If yes, how many instalments did you repay so far? 1...........2.......

13.3 If no, What is the reasons?

(a) Lack of suitable/adequate staff for the purpose of efficient supervision and recovery of loan.

(b) Clearing old debt to money lenders/commission agent / relatives

(c) Expenditure on consumption.

(d) Unfavourable price structure.

14.0 Do you feel that your PACB is a successful one? Yes/No.

14.1 If yes, what reasons you assign for the success?

a. b. c.

14.2 If no, what reasons you assign for the failure?

a. b. c.

15.0 Reasons for default of repayment of loan?

Sl. No	*Particulars*
(a)	Loss is operation
(b)	Marketing problem
(c)	Delay in sanctioning the Loan
(d)	Natural calamities like drought/floods
(e)	Any heavy interest rate
(f)	Low yield per acre
(g)	Inadequate level of irrigation facilities
(h)	Inadequate level of loan amount
(i)	Misutilisation of loan
(j)	No transport facilities
(k)	Any others (specify)

16.0 Any suggestions about co-operative bank?

a. b. c.

Place :

Date :

Signature

BIBLIOGRAPHY

REPORTS

Annual Credit Plan, 1999-2000, Vellore District, Indian Bank, Vellore.

Annual Credit Plan, 2000-2001, Vellore District, Indian Bank, Vellore.

Annual Credit Plan, 2001-2002, Vellore District, Indian Bank, Vellore.

Annual Report 2001-2002, Supplement to RBI Bulletin, Mumbai, September, 2002.

Cox Arther, F., Gazetteer of North Arcot District, Madras, 1985.

Economic Survey 1994-95, Ministry of Finance, Government of India.

Economic Survey 2000-2001, Ministry of Finance, Government of India.

Economic Survey 2001-2002, Ministry of Finance, Government of India.

Government of India, Census Hand Book of North Arcot District, Village and Town Directory, 1961.

Government of India, Census Hand Book of North Arcot District, Village and Town Directory, 1971.

Government of India, Census Hand Book of North Arcot District, Village and Town Directory, 1981.

Government of India, Census Hand Book of North Arcot District, Village and Town Directory, 1991.

Government of India, Census of India, Primary Census Abstract, 1961.

Government of India, Census of India, Primary Census Abstract, 1971.

Government of India, Census of India, Primary Census Abstract, 1981.

Government of India, Census of India, Primary Census Abstract, 1991.

Government of India, Census of India, Series 34, Provisional Population Table, Directorate of Census, 2001.

Government of Madras, Manual of Administation of Madras Presidency, Madras, 1985.

Hand Book of Indian Agriculture, Vikas Singhal, Vikas Publishing House Pvt. Ltd., New Delhi.1995.

Hand Book of Statistics on Indian Economy, RBI, Mumbai, 1999.

KHUSRO, A.M., (Chairman), "A Review of the Agricultural Credit System in India", RBI, 1989.

NARASIMHAM, M., (Chairman), "Report of the Committee on the Financial System in India, RBI, 1991.

Potential linked Credit Plan, Vellore District 2000-2001, NABARD, Regional Office, Chennai.

Potential linked Credit Plan, Vellore District 2001-2002, NABARD, Regional Office, Chennai.

Potential linked Credit Plan, Vellore District 2002-2003, NABARD, Regional Office, Chennai.

Proceedings of the Fifteenth and Sixteenth meetings of the Agricultural Credit Board, RBI, 1981.

Report on Currency and Finance (1997-98), Vol.I, Economic Review, RBI.

Report on Trend and Progress of Banking in India 1998-99, Supplement to RBI Bulletin, December, 1999.

Report on Trend and Progress of Banking in India 1999-2000, RBI December 2000.

Report on Trend and Progress of Banking in India 1999-2000 supplement to RBI Bulletin, December, 2000.

Statististical Statements Relating to the Co-operative Movement in India, 1995-96, Part-I, NABARD, Mumbai.

Statistical Hand Book-2000, Department of Economics and Statistics Government of Tamil Nadu, Chennai, 2001.

Statistical Outline of India, 1999-2000, TATA Service Limited, Mumbai, 1999.

Tamil Nadu - An Economic Appraisal 1991-92, Department of Evaluation and Applied Research (DEAR), Government of Tamil Nadu.

Tamil Nadu - An Economic Appraisal 1993-94, Department of Evaluation and Applied Research (DEAR), Government of Tamil Nadu.

Tamil Nadu - An Economic Appraisal 1996-97, Department of Evaluation and Applied Research (DEAR), Government of Tamil Nadu.

Tamil Nadu - An Economic Appraisal 1997-98, Department of Evaluation and Applied Research (DEAR), Government of Tamil Nadu.

The Hindu, Survey of Indian Industry - 2003.

94th Annual Reports and Statement of Accounts, 1999-2000, The Tamil Nadu State Apex Co-operative Bank Ltd., Chennai.

JOURNALS

Adinarayana, S. and Lakshminarayan. M, "Impact of Institutional Production Credit on farm Productivity in Kasimkota panchayat Samithi of Visakhapatnam District (A.P)", *Indian Journal of Agricultural Economics,* Vol.XLIII, No.3, July-September, 1988.

Alamelumangai, R., "Can Co-operatives Survive and succeed in the context of Economic liberalisation? *The Tamil Nadu Journal of Co-operation,* Vol.87, No.5, August, 1995.

Amrik Singh Saini, Raj Vir Singh and PATEL, K.K., "Rural Credit Management through integrated crop and milk production", *Indian Journal of Agricultural Economics,* Vol.XLIII, No.3, July-September, 1988.

Anandharajakumar, P. and Arockiaraj, K. "Primary Agricultural Credit Socieities Instrument for Rural Development", *The Tamil Nadu Journal of Co-operation,* Vol.87, No.8, November, 1995.

Atchi Reddy, M. and Prabhākar, "Rural Credit in Karim Nagar District", *Indian Journal of Agricultural Economics,* Vol.XLIII, No.3, July-September, 1988.

Balishter and Chauhan T.R., " A study of Co-operative Finance in Agriculture". *Indian Co-operative Review,* Vol.XIX, No.3, April, 1982.

Beohar and Khare, "A study of the Recovery and overdues of Agricultural loan issued by Bishnoo Datta Central Co-operative Bank, Branch Umaripan, Jabalpur district (M.P)", *Indian Journal of Agricultural Economics,* Vol.XLIII, No.3, July-September, 1988.

Bhagawan Rao and Kanakaiah, "Financing of Borrowers by Farmers Service Co-operative Societies - A case study", *Indian Co-operative Review,* Vol.XXI, No.1, July, 1983.

Bhalerao, M.M., Malati and Kavidra Singh, "Regional inequalities in long-term institutional credit -A study of land Development Bank", *Indian Journal of Agricultural Economics,* Vol.XLIII, No.3, July-September, 1988.

Bhosal, S.K. and Bangat, S.B, "A Study into the overdues of Co-operative Loans in Maharashtra", *Indian Journal of Agricultural Economics,* Vol.XLIII, No.3, July-September, 1988.

Chandra Prasad, J. and Visweswara Rao, K.V.V.S., "Co-operative Credit Agencies: An Evaluation", *Kurukshetra,* Vol.46, No.11, August, 1998.

Chauhan, T.R., Balishter and Rajkumar, "Availability and Extent of Utilisation of Credit in Agriculture," *Indian Co-operative Review,* Vol.XX, No.1, July, 1982.

Datta, T.N., "Analysis of Default in Organised Sector Credit - A study of minor Irrigation scheme in West Bengal," *Indian Journal of Agricultural Economics,* Vol.XLIII, No.3, July-September, 1988.

Dinesh, K. Marothia, "Production Credit Management in Rice Farming System of Chhattisgarh Region", *Indian Journal of Agricultural Economics,* Vol. XLIII, No. 3, July-September, 1988.

Duhan, T.C. and Singh, "Institutional Credit for Dairying in Haryana State", *Indian Journal of Agricultural Economics,* Vol.XLIII, No.3, July-September, 1988.

Ganesh, P., Gupta and Sudha Ran Sahu, "Strategy for promotion of loyalty in the Co-operative System", *Co-operative perspective, Journal of Co-operative Management,* Vol. 34, No. 4, Jan-March, 2000.

Gangadharan, T.P., "Intirmities in Rural Credit Policy-A case study of Farmers Loans in Kozhikode District", *Indian Journal of Agricultural Economics,* Vol.XLIII, No.3, July-September, 1988.

Gangwar, A.L. and Kusam Aggarwal, "Borrowing and Repayment Performance of Farmers Pertaining to Institutional Loans in Kurukshetra District, Haryana", *Indian Journal of Agricultural Economics,* Vol. XLIII, No. 3, July-September, 1988.

Gayal, S.K. and Pandey, "Trend Analysis of Primary Agricultural Co-operative Credit and Service Society in Haryana," *Indian Co-operative Review,* Vol. XXVI, No. 1, July, 1988.

Gulab Singh Azad, "Role of Co-operatives in Rural Development," *The Tamil Nadu Journal of Co-operation,* Vol.88, No.11, February, 1997.

Gupta and Verma, "Role of Co-operative Credit in Increasing Agricultural Production in M.P", *Indian Co-operative Review,* Vol. XXII, No. 1, 1986.

Hanumanthaiah, C.V., Raju, V.T. and Naidu, M.R., "Temporal Variation in Economic and financing Aspects of cotton farms-A case study", *Indian Journal Agricultural Economics,* Vol.XLIII, No.3, July-September, 1988.

Hemchand Jain, "Impact of Institutional Credit on Tribal and Non-tribal Farms in Mandla District of Madhya Pradesh", *Indian Journal of Agricultural Economics,* Vol.XLIII, No.3, July-September, 1988.

Julka, A.c., Soni, R.N. and Arora, S., "Profile and cost of Rural Credit: A case study", *Indian Journal of Agricultural Economics,* Vol.XLIII, No.3, July-September, 1988.

Kaliya Perumal, P. and Loganathan P., "Agricultural Credit in Elakurichi Village a Micro Analysis", *Indian Co-operative Review,* Vol. XVII, No. 2, October, 1989.

Kalyan Bhattacharya, "Financial intermediation and Rural Development: Role of Co-operative Credit in Agricultural Development", *The Indian Economic Journal,* Special Conference number for 1978, Part-II, Vol.26, No.2, October-December, 1978.

Kamta Prasad, "Financial Intermediaries and Rural Development", The *Indian Economic Journal,* Special conference number for 1978, Part-II, Vol. 26, No. 2, October-December, 1978.

Lal R.C., and Lavania R.P., "Impact of Co-operative Credit on Agriculural Production and Income", *Indian Co-operative Review,* Vol. XXIII, No. 3, January, 1986.

Mitra, A. and Sain, K., "Financing of Agriculture by Co-operative Banks in Burdwan district", *Indian Co-operative Review,* Vol. XX, No.1, July, 1982.

Mudgal, M.K., "Co-operatives at Cross-Roads", *Kurukshetra,* Vol.46, No.8, May, 1998.

Munidoraswamy Naidu, K. and Surendrababu, K.B., "Rural Credit Policy: Need for Transcending Credit Needs by creating debt Capacity", *Indian Journal of Agricultural Economics,* Vol.XLIII, No.3, July-September, 1988.

Pandey, R.K. and Guglani, P.L., "Trends in Recovery overdues and outstanding position of Agricultural Loans, Ranch District (Bihar)", *Indian Journal of Agricultural Economics,* Vol.XLIII, No.3, July-September, 1985.

Pandey, R.K. and Ashok Kumar, "Study of Growth and Disparity in Flow of Co-operative Credit in India," *Indian Journal of Agricultural Economics,* Vol.XLIII, No.3, July-September, 1988.

Radhakrishnan, V. and Mukundan, "Supply and Utilisation of Short-term Co-operative Agricultural Credit in Palghat district", *Indian Journal of Agricultural Economics,* Vol.XLIII, No.3, July-September, 1988.

Raghu Ram, P. and Bhavani Devi, I,"An analysis of Tangible means to Reduce Incidence of Default in Commercial Bank Lendings," *Indian Journal of Agricultural Economics,* Vol.XLIII, No.3, July-September, 1988.

Rajkishore Panda, "Credit Finance of Primary Agricultural Co-operative Societies Among Farm Households in Orissa", *Indian Journal of Agricultural Economics,* Vol.XLIII, No.3, July-September, 1988.

Rajput, S.S., "Rural Credit and its cost - A Case study", *Indian Co-operative Review,* Vol.XXIII, No. 3, January, 1986.

Ramachandra Bhat and Aruna Rao, "The Evaluation of the Growth of Primary Agricultural Credit Co-operative Societies in Karnataka 1965-1978", *Indian Co-operative Review,* Vol.XIX, No.3, April, 1982.

Rangarajan, C. "Issues in Rural Credit", *The Tamil Nadu Journal of Co-operation*, Vol. 89, No. 5, August, 1997.

Ratnaparkhe, A.N. and Pagire, B.U., "A study into the Appropriateness in the scale of Finance by the Co-operatives in Maharashtra State," *Indian Journal of Agricultural Economics*, Vol. XLIII, No.3, July-September, 1988.

Rayudu, C.S., "Participation of Co-operatives in Small farmers Development -A case study", *Indian Co-operative Review*, Vol.XXI, No.4, April, 1984.

Sankaraiah, G. and Munidoraiswamy Naidu, K., "Financing Small farmers by Co-operatives", *Indian Co-operative Review*, Vol. XXI, No.1, July, 1983.

Sarker, S.C.., Mitra, A and Mukhergee, A.K., "Institutional Credit in Developed and Underdeveloped Villages of Naidu District West Bengal an appraisal," *Indian Co-operative Review*, Vol. XXIV, No. 2, October, 1986.

Satish Verma and Vijay Chopra, "Demand for Bank Credit in the Agricultural sector in India", *Indian Journal of Agricultural Economics*, Vol. XLIII, No. 3, July-September, 1988.

Satyanarayana, I, Ch. Rushyendra Rao and Venkatesa Rao, "Subsidy to Identified Beneficiaries by FSS-A case Study," *Indian Co-operative Review*, Vol.XXIII, No.3, January, 1986.

Satyasai, K.J.S. and Badatya, K.C., Restructuring Rural Credit Co-operative Institutions, *Economic and Political Weekly* Vol. XXXV, No. 5 Jan 29, Feb 4, 2000.

Sharma, M.L, and Teweri, "Effect of Alternative Production Credit Allocation Policies on growth and Equity in a Community Development Block", *Indian Journal of Agricultural Economics*, Vol.XLIII, No.3, July-September, 1988.

Singh, S.K., Ram Iqbal Singh and Singh, G.N., "Impact of Rural Credit through Co-operative in Agriculural Development in Eastern Uttarpradesh," *Indian Journal of Agricultural Economics*, Vol. XLIII, No.3, July-September, 1980.

Singha, R.P. and Sharma, B.M., "Role of Credit and Technology in Increasing farm Income Labour Employment - A case Study", *Economic Affair*, Vol.33, quarterly, March, 1988.

Singhal, C.D., "Evaluation of Role of Primary Credit and Service Societies (PACS) in Haryana-A case study," *Indian Co-operative Review*, Vol. XXVI, No.1, July, 1988.

Srinivasan, S., "Long-Term Credit in Anna District Tamil Nadu-A case study", *Indian Journal of Agricultural Economics*, Vol.XLIII, No.3, July-September, 1988.

Srivastava, G.C. and Lal, K.K., "Economic Rationale of Credit use on small Farms - A case study in Bihar", *Indian Journal of Agricultural Economics*, Vol.XLIII, No.3, July-September, 1988.

Suryawanshi, S.P., "Credit Requirement and its Gap on the Farm in Bhima Command Area of Western Maharashtra," *Indian Journal of Agricultural Economics*, Vol. XLIII, No. 3, July-September, 1988.

Tamil Mani, B. and Manivel, S, "Repaying the Dues to the Co-operative Banks is the best culture for the growth of Co-operatives," *The Tamil Nadu Journal of Co-operation*, Vol.89, No.7, October, 1997.

Vaidya, C.S, Sikka,B.K. and Swarup, R, "Cost of Credit in Hill Agriculture: A Spatial Analysis," *Indian Journal of Agricultural Economics*, Vol. XLIII, No. 3, July-September, 1988.

Vaikunthe, L.D., "Utilisation of Credit in Agriculture - A case study", *Indian Co-operative Review*, Vol. XXV, No. 4, April, 1988.

Verma, S.K, "Rural Development through Co-operatives", *Kurukshetra*, Vol. 47, No. 8, May, 1999.

Viswanadha Rao and Mohandoss, "Co-operative credit certain Dimension and problems", *Indian Co-operative Reivew*, Vol. XXIII, No. 3, January, 1986.

THESIS

Govindarajan, P. Dr., 'Economics of Poverty, Rural Poverty an Institutional Approach, with reference to North Arcot District', Ph.D. Thesis, University of Madras, Madras, 1984.

Sivaprakasam, R. 'Housing conditions of scheduled caste in Vellore District (Tamil Nadu) : A study', Ph.D. Thesis, University of Madras, Chennai, 2000.

Radhakrishnan, N. "Institutional Credit for Agricultural Sector - A case study of Mambakkam Primary Agricultural Co-operative Credit Society: 1985-86 to 1992-93, M.Phil., Thesis, Department of Economics, University of Madras, 1994.

WEBS

Government of Tamil Nadu, Website: www.tn.gov.in

BOOKS

Baidyanath Misra, *Co-operative movement in India*, A.P.H. Publishing Corporation, New Delhi, 1997.

Bansil, P.C., *Agricultural Problems of India*, Oxford and IBH Publishing Co, New Delhi, 1981.

Bedi, R.D., *Theory, History and Practice of Co-operation*, International Publishing Hosue, Meerut, 1987.

Bedi, R.D., *Theory, History and Practice of Co-operation*, Loyal Book Depot, Meerut, 1986-87.

Bhatia, B.S., Verma, H.L. and Mahesh C. Garg, *Encyclopaedia of Co-operative Management, Co-operative Banking: Leavers of Rural Economy*, Vol.3, Deep and Deep Publications, New Delhi, 1994.

Co-operation in India and Abroad, Tej Bahadur Singh Chandel, Kanpur, 1968.

Clarence Zuvekas, *Economic Development - An Introduction*, The Macmillan Press Ltd., London, 1979.

Cohen, R.L., *The Economics of Agriculture*, Cambride University Press, London, 1965.

Dandekar, M.M., *The Indian Economy 1947-1992, Agriculture*, Vol.I, Sage Publication, India Pvt. Ltd., New Delhi, 1994.

Dutta, S.K., *Co-operative Societies and Rural Development*, Mittal Publications, New Delhi, 1991.

Eleanor, M., Hough, *The Co-operative Movement in India*, Oxford University Press, 1966.

Farmer, B.H., *Green Revolution? Technology and change in Rice-Growing Areas of Tamil Nadu and Srilanka*, The Macmillan Press Ltd., London, 1980.

Ghosal, S.N., *Agricultural Financing in India*, Asia Publishing House, Bombay, 1929.

Gopal Lal Jain, *Rural Development*, Mangal Deep Publications, Jaipur, 1997.

Jain, P.C., *Economic Problems of India*, Chaitanya Publishing House Allahabad, 1965.

Jaya S. Anand, *Co-operative Agricultural and Rural Development Banks*, Atlantic Publishers and Distributors, New Delhi, 1999.

Krishna Kumari, A.B., *Co-operative Banking for Agricultural Development*, Deep and Deep Publications, New Delhi, 1998.

Kulkarni, K.R., *Theory and Practice of Co-operation in India and Abroad*, Vol.I. Co-operators Book Depot, Bombay, 1968.

Kulkarni, K.R., *Theory and Practice of Co-operation in India and Abroad*, Vol.II (Part II), Co-operators Book Depot, Bombay, 1965.

Mamoria, C.B., *Agricultural Problems of India*, Kitab Mahal Pvt. Ltd., Allahabad, 1995.

Mamoria, C.B. and Saksena, R.D. *Co-operation in India*, Kitab Mahal, Allahabad, 1977.

Maslennikov, V., *The Co-operative Movement in Asia and Africa,* Progress Publishers, Moscow, 1983.

Mathur, B.S. *Co-operation in India,* Sahitya Bhawan, Agra, 1988.

Nakkiran, S., *Co-operative Banking in India,* Rainbow Publications, Coimbatore, 1980.

Palaniappa Mudaliar, A., *The Co-operative Movement in Madras State-A Brief Survey,* The Tamil Nadu Co-operative Union, Madras, 1958.

Raj Kishor Pany, *Institutional Credit for Agriculture in India,* Ashish Publishing House, New Delhi, 1985.

Ramasamy, C., Hazell, B.R., Peter and Aiyasamy, P.K., North Arcot and Green Revolution, Ch.2, In Peter B.R.Hazell and Ramasamy, C., (eds) *The Green Revolution Reconsidered, The Impact of High Yielding Rice varieties in South India,* The International Food Policy Research Institute, Oxford University Press, Oxford, 1991.

Ranga Reddy, A., *Agricultural Development Rural Credit and Problems of its Recovery,* Mittal Publications, New Delhi, 1990.

Sami Uddin and Mahfoozur Rahman, *Co-operative Sector in India-After Independence,* S. Chand and Company Ltd., New Delhi, 1983.

Sivasubramanian, A.P., *The practive of Co-operative Banking,* Arul Selvi Publications, Tirunelveli, 1986.

Suresh Reddy, Y. and Reddeppa Reddy, A., *Co-operatives and Rural Development,* Anmol Publications Pvt. Ltd., New Delhi, 1997.

Tandon, B.B and Kulwinder Kaur, *Indian Economic Problems,* S.Chand and Company (Pvt) Ltd., New Delhi, 1987.

Thirunarayanan, R., *Co-operative Banking in India,* Mittal Publications, New Delhi, 1996.

Tyagi, B.P. *Agricultural Economics and Rural Development,* Jai Prakash Nath & Co, Meerut, 1990.

Tyagi, R.B, *Recent Trends in the Co-operative Movement in India,* Asia Publishing House, Bombay, 1968.

Udduman Mohideen, K.S.S., *Institutional Credit for Agricultural Development,* Mittal Publications, New Delhi, 1991.

Vasant Desai, *A Study of Rural Economics-A Systems Approach,* Himalaya Publishing House, Bombay, 1983.

Victor, M.A., *Co-operation in Madras State,* Palaniappa Brothers, Madras, 1964.

